Navigating the FDA Regulatory Processes

This book provides a unique knowledge source of the FDA's regulatory processes to anyone interested in the development and approval of biomedical products.

Navigating the FDA Regulatory Processes will be of particular importance for those innovating in biomedical devices, pharmacological therapies, and combination products, who are interested in considering regulatory affairs early on in product development. The authors offer a comprehensive exploration of FDA regulations, providing industry insights and guidance for navigating the complexities of drug and device development. The book offers concrete examples, industry perspectives, and historical context, ensuring clarity and aiding in successful understanding of the requirements for regulatory compliance. Readers will also gain a thorough understanding of FDA regulatory processes, from securing approvals to navigating expedited pathways. Lastly, the book provides unique insights into the FDA approach for emergency preparedness, including discussing the recent COVID-19 pandemic.

Biomedical industry personnel, engineers, scientists, entrepreneurs, medical doctors, students, and anyone interested in the development and approval of medical products in the US will find this book useful and essential.

Navigating the FDA Regulatory Processes

An Integrated Clinical and Engineering Approach

Edited by

Pablo Gurman, Yitzhak Rosen, and Noel M. Elman

CRC Press
Taylor & Francis Group
Boca Raton London New York

CRC Press is an imprint of the
Taylor & Francis Group, an **informa** business

Designed cover image: Sonia Minetti

First edition published 2025
by CRC Press
2385 NW Executive Center Drive, Suite 320, Boca Raton FL 33431

and by CRC Press
4 Park Square, Milton Park, Abingdon, Oxon, OX14 4RN

CRC Press is an imprint of Taylor & Francis Group, LLC

ISBN: 978-1-4822-0708-8 (hbk)
ISBN: 978-1-0329-0474-0 (pbk)
ISBN: 978-1-3151-1791-1 (ebk)

DOI: 10.1201/9781315117911

Typeset in Times
by Deanta Global Publishing Services, Chennai, India

Contents

About the Editors

Dr. Pablo Gurman, MD, MSM, received his MD from Buenos Aires University School of Medicine and a master's degree in management from Babson College. His experience spans 15 years of experience in R&D in the biotech and the pharmaceutical industry, as well as more than 15 years of R&D in microsystems and nanotechnology working at MIT, The University of Texas at Dallas and Argonne National Laboratory (Artificial Retina Program). His primary interests include drug development, medical device development with focus on microsystems and nanotechnology, biotechnology and public health.

Dr. Yitzhak Rosen, MD, is a graduate of Tel Aviv University of Medicine. He is a practicing physician. Dr. Rosen was affiliated with the Cardiology Department at SUNY Downstate Medical Center, New York and served as a visiting research scientist at the Institute for Soldier Nanotechnologies at MIT. Currently, he is the president and CEO of Superior NanoBioSystems LLC, a biomedical company. Dr. Rosen has also served as a medical officer and physician in militarily active areas with the Israel Defense Forces (IDF). He completed his medical internship at Rabin Medical Center and has worked at the Oncology Institutes of both Rabin and Sheba Medical Centers in Israel. Notably, he invented a microfluidic chip platform, funded by DARPA, for rapid blood typing and cross-matching in mass casualty situations, in collaboration with the MEMS and Nanotechnology Exchange. Additionally, he has developed several medical ultrasound technologies.

Noel M. Elman, PhD, is the CEO and founder of GearJump Technologies, LLC, a company focused on creating biotechnological solutions for public health. He is also a faculty member at the Buenos Aires Institute of Technology. Previously, Dr. Elman worked at Draper Laboratory, an MIT-affiliated and DoD-supported R&D center, where he was a distinguished member of the technical staff. Before that, he served as a research scientist and principal investigator at the Institute for Soldier Nanotechnologies at MIT, leading a translational research group dedicated to developing biotech, biomedical, and public health technologies. Dr. Elman's research and development focus is on rapid translation from idea conceptualization to product realization. He holds bachelor's and master's degrees in electrical engineering from Cornell University and a PhD in physical electronics from Tel Aviv University. Additionally, he completed postdoctoral studies at MIT, where he researched various micro- and nanodevices for therapeutics and diagnostics.

Contributors

Harvey M. Arbit – Founder and Consulting Partner, Arbit Consulting, LLC and Adjunct Professor, Albany College of Pharmacy and Health Sciences; Adjunct Professor, College of Pharmacy, University of Minnesota

Angela Bellos – Senior Manager Regulatory Affairs, CMC; Resilience US, Inc.

Oliver V. Cácananta – Owner and Principal Consultant, OVC Insights LLC, California, USA and Senior Program Manager, Head of Regulatory Affairs for UCSF Investigational Cellular Therapy, California, USA

Shuo Chen – Senior Manager, Global Regulatory Affairs Oncology, at Regeneron Pharmaceuticals

Anurag Patel – Director, Regulatory Affairs-CMC (Medical Devices and Combination Products), Merck

Giridhar Prasad – Senior Manager Global Regulatory Affairs Biologics, Sanofi

Jack Russell – Senior Manager Regulatory Affairs, CMC; Resilience US, Inc.

Evan B. Siegel – President and CEO, Ground Zero Pharmaceuticals, Inc., Irvine, CA and Adjunct Professor, University of Queensland, Brisbane, Australia

Doris Snow – Senior Manager Regulatory Affairs, CMC; Resilience US, Inc.

Wrenda Teeple – President, Arbit Consulting, LLC

Raymond Vennare – Chairman & CEO Predictive Oncology, Inc., USA

Foreword

Due to the complexity involved with developing new and emerging innovative drug and device products for American consumers, the critical regulatory role of the Food and Drug Administration (FDA) grows ever more important in protecting public health via continued evaluation of the safety, efficacy, and security of human and veterinary drugs, biological products, medical devices, our nation's food supply, cosmetics, and products that emit radiation in the United States. It cannot be understated that navigating the labyrinthine corridors of this regulatory agency will require deciphering complex terminology, and understanding the dynamic dance between industry and regulators can be a daunting task.

Therefore, whether you are a student, industry professional, or simply someone who is curious and interested in learning more about the FDA's regulatory role, the following ten chapters of this book will provide a detailed compilation of industry insights and perspectives on the regulation of biomedical products.

Chapter 1, an introductory chapter, provides an outline of its contents, aiding readers in navigating the complex subjects covered within this book.

Chapter 2 showcases the vital interactions between industry and the FDA, emphasizing the significance of managing expectations and adherence to timelines in the context of commercial biomedical endeavors.

Chapter 3 dives into the extensive process of securing FDA approval for new drug applications, complemented by concrete examples and insights.

Chapter 4 offers a concise summary of FDA regulatory pathways pertinent to generic drugs, with a historical perspective that underscores the transformative impact of regulations on pharmaceutical accessibility.

Chapter 5 further explains the regulatory requirements governing medical devices, categorizing them based on risk classifications, and presents pathways to market.

Chapter 6 provides an extensive overview of the regulatory landscape for biologics, embracing the groundbreaking advancements in biotechnology and the diverse range of products they encompass.

Chapter 7 navigates the complexities of combination products, offering clarity on the regulatory intricacies surrounding these multifaceted innovations.

Chapter 8 provides a pragmatic survey of FDA expedited pathways, a critical resource for those seeking to streamline product approvals, especially in times of urgency.

Chapter 9 examines investigator-initiated sponsored research, emphasizing the importance of transparency and adherence to regulations for the successful execution of clinical studies.

Finally, Chapter 10 delves into the domain of medical countermeasures, revealing the regulatory considerations during public health emergencies, with a particular focus on the lessons drawn from the recent COVID-19 pandemic.

With deep appreciation for the author's commitment to clarity and precision, readers will find a wealth of knowledge, industry insights, and guidance presented in an objective and accessible manner. From the introductory chapter, which provides a comprehensive overview of the book's contents, to the deep dives into the regulation of drugs, generics, medical devices, biologics, combination products, expedited pathways, investigator-initiated research, and medical countermeasures, this book leaves no stone unturned.

1 Introduction

Pablo Gurman, Yitzhak Rosen, and Noel M. Elman

CHAPTER 1: INTRODUCTORY CHAPTER

This introductory chapter provides a summary of each chapter, allowing the reader an easy navigation and understanding of this book. The overall intended purpose is to provide for the reader a quick guide that provides an overview of the key subjects covered and thus facilitates the overall absorption of the expansive material developed within each chapter, which will be of interest to anyone who is trying to understand the regulation of biomedical products in the US. This is relevant not only to professionals working in the regulatory departments of pharmaceutical, medical, or biotech companies but also to a broader audience, including researchers in the biomedical field. Thus, this book should be thought of as a source collection of expert opinions dedicated to enhancing the understanding of the FDA regulation process for biomedical products.

CHAPTER 2: DANCING WITH THE FDA: MANAGING EXPECTATIONS

The second chapter provides a segue to understanding the implications of industry-FDA interactions. The author offers a detailed description of the interactions that are crucial for managing expectations and uncertainty during the medical product development process. The concept of the dance between the FDA and the sponsor seeking approval is ingeniously described in terms of the timing of the approach to the Agency as well as the timeline for the interactions. Time is the key factor that will determine the fate of the project, as it is the key parameter for a commercial entity to be able to sustain itself. The longer the time to reach a milestone, the more complex, challenging, and onerous the process will be. This chapter describes the critical junctures that entrepreneurs and companies face when dealing with a marketing application, such as for a medical device, e.g., Class I, II, or III. This chapter emphasizes the interactive understanding between regulatory requirements and their impact on the assessment of the overall business opportunity, in terms of strategy, use of resources, financial assumptions, and overall market potential. Therefore, this chapter is an invaluable short guide for the researcher or entrepreneur who is attempting to venture into a new commercial project that requires interactions with the FDA. The chapter provides a reference to manage the expectations that operate in different dimensions but still interact with one another. The chapter describes the greatest challenges ("hurdles") with regard to the commercialization of medical technology, including regulatory aspects, reimbursement, intellectual property, and market opportunity. These important considerations need to be addressed as early as possible when the project is being defined, especially for the entrepreneur who is

DOI: 10.1201/9781315117911-1

starting the process for the first time and needs to define a timeline and a viable commercial pathway to success. The chapter provides a unique insight into the process, starting from proof of concept to commercialization, which still relies on scientific and engineering concepts but inexorably depends, in the final estimation, on the regulatory pathway chosen. This chapter therefore presents a unique perspective on the FDA pathways and the required interaction with the sponsor seeking approval, licensing, or clearance, as well as the management of expectations to develop a realistic timeline.

CHAPTER 3: NEW DRUG APPLICATION

This chapter provides a comprehensive review of the process for a New Drug Application (NDA). The author provides a blueprint of the multiple steps required to obtain FDA approval for new drug applications. This chapter also provides specific examples and references to provide concrete evidence of these processes. Definitions and descriptions of various types of drug entities are provided, including New Molecular Entity, New Active Ingredient, New Dosage Form, and New Combination or New Ratio of Multi-API Drugs, New Formulations, New Indications, products previously marketed but without an approved NDA, Prescription to Over-the-Counter Switch, and New Indications or Claims.

In addition, the process flow describing the steps and mechanics to engage with the FDA is described in detail, including requirements for preclinical and clinical studies. The author describes the importance of early meetings with the FDA to assess the tasks and deliverables required to complete an application. The process is thus described in detail, including the definition of Investigational New Drug Applications, as well as descriptions of clinical study phases (Phase 1–4). Finally, a brief description of NDAs for pediatric applications is also provided.

This chapter is therefore highly recommended for the new researcher, entrepreneur, or even medical doctor, or patient who wishes to become acquainted with the intensive process required to introduce a new drug into the public domain. By the end of this chapter, readers will become familiarized with the multiple steps, processes, terms, and acronyms commonly applied in the industry. This chapter should be used as an invaluable reference source that provides an initial guide to the FDA process for reviewing and approving an NDA.

CHAPTER 4: BASICS OF GENERIC DRUG REGULATION

This chapter provides a succinct summary of the FDA regulatory pathways related to generic drugs, including a brief historical overview of the evolution of the FDA's regulations for generic drugs. The term "generic" drugs generally refers to small-molecule drug products that have the same active ingredients (AIs), dosage form, strength, indication, and route of administration as a brand-name drug (innovator). When the brand-name drug patent has expired, the FDA can determine whether the generic drug could be a substitute for the brand-name drug and whether it could convey the same clinical benefit to the patient. Generic drugs account for approximately

90% of all the prescriptions filled in the US. Interestingly, the chapter provides the initial rationale for congressional passage of the Hatch-Waxman Act in 1984, which was to provide incentives for generic drug companies to overcome challenges presented by products whose patents have expired, decrease development costs, and accelerate development cycles. These drugs then become more readily available for the general population. The Act provides marketing exclusivity to the innovator products, irrespective of their patent expiration, to reward them for the time, costs, and resources attendant to the initial development of these drug products. Overall, the approval process of generic drugs via Abbreviated New Drug Applications (ANDAs) is less complicated than that for an innovator New Drug Application (NDA), thus providing a faster pathway to commercialization and facilitating access to medicines by the general population in a timely manner. The ANDA does not require submission of clinical data to support safety and efficacy since the data should have been already provided by the original innovator with the initial NDA. A bioequivalence or bioavailability study in healthy human volunteers is required. A generic drug must be therapeutically equivalent to the branded product. It must meet the same requirements for identity, strength, purity, and quality as the innovator. The generic drug must be pharmaceutically equivalent as well as bioequivalent. The ANDA needs to include three key elements related to the Reference Listed Drug (RLD): bioequivalence data, quality data, and administrative information related to labeling and patent information. Finally, the chapter summarizes the Generic Drug User Fee Act (GDUFA), which is a law, passed on October 1, 2012, designed to update laws on generic drugs, expedite access to safe and effective generic drugs to the public, and reduce costs to the industry. Overall, this chapter provides a practical guide to regulatory requirements related to generic drugs, with important considerations to take into account from the early development phase through commercialization.

CHAPTER 5: INTRODUCTION TO MEDICAL DEVICE REGULATION

This chapter provides an informative review of regulatory requirements related to medical devices. A short and illustrative overview of historical references is provided, as well as the initial need to define regulatory requirements to manage the safety of medical devices. The laws designed to regulate medical devices were passed relatively recently, in 1976, under the Medical Device Amendments to the Food, Drugs and Cosmetics Act. The Center for Devices and Radiological Health (CDRH) is the FDA's administrative body responsible for regulating medical devices and radiation-emitting electronic products intended for both medical and non-medical purposes. Medical devices are classified according to different risk classifications. Class I involves the lowest risk devices that involve minimal potential harm and require only general controls for assessing safety and effectiveness. The regulatory pathway to commercialization is a 510(k), *de novo* 510(k) exempt application. Class II involves moderate risk devices that may require special controls for assessing safety and effectiveness. The regulatory pathway also involves the 510(k) application. Class III involves the highest risk devices that involve Premarket Approval (PMA) applications with more comprehensive controls for assessing safety and effectiveness, and the need for conducting clinical trials prior to submission of the marketing application under Investigational Device Exemptions (essentially INDs for devices). The chapter provides several examples for each of these categories, as well as a quick summary of requirements for manufacturers that need to follow current good manufacturing practice (GMP) as well as quality systems requirements (QSR). In addition, the chapter provides a practical review of medical regulatory requirements for device accessories as add-ons requirements to complement, support, supplement, or even augment performance of a device. The concept of multifunctional device products and convenience kits is also introduced to illustrate the FDA's oversight on the regulatory requirements for medical devices that have multiple functionalities or are combined to achieve a specific function. Finally, multiple regulatory pathways are briefly described to provide a quick summary of the requirements. These pathways include: 510(k) Premarket Notification, PMA, De Novo Classification Requests, and Humanitarian Device Exemption (HDE). The latter is granted to a sponsor through the Office of Orphan Products Development (OOPD). The Breakthrough Device Program is also described, for which two criteria must be met. The first is that the device must address a life-threatening or irreversibly debilitating condition. The second is that devices need to be considered breakthrough technologies, without any existing alternatives, offer advantages over existing products, or be in the best interest of the patients vis-à-vis risk-benefit.

The chapter includes additionally important topics such as unique device identifiers needed to trace devices from manufacture to distribution, device quality systems, and medical device safety.

Finally, the chapter includes a discussion of digital health products, a topic that is gaining increasing importance and impacting several therapeutic areas. The chapter describes the scope of digital health, including mobile applications and health information. Importantly, the chapter defines the conditions under which digital functionalities are exempt from FDA regulation.

CHAPTER 6: FDA AND REGULATION OF BIOLOGICS

This chapter provides a comprehensive review of regulations for biological products, otherwise known as biologics. The chapter starts with a clear definition of biological products combined with a brief historical background. It is important to note that the FDA administrative body with jurisdiction over biologics is the Center for Biologics Evaluation and Research (CBER). Biologics represent one of the most important next-generation breakthroughs in biotechnology, through which the emergency-use approval of ribonucleic acid (RNA)-based vaccines has been used against the COVID-19 pandemic. There is a wide variety of biological products, which are typically described as large and complex molecules usually derived from biological systems such as the fermentation of bacteria, yeast, and cell culture. The definition of biologics, therefore, covers a number of products, including vaccines, cells and tissues for regenerative medicine, tissue transplantation, and gene therapy and engineering. Blood products and allergenics are also considered biologics, as are nucleic acids and most proteins. Combination products that integrate biologics with devices or drugs also represent the next generation of therapeutics and diagnostics. Product development for biologics, including discovery, preclinical, clinical development, and commercialization, is clearly summarized, providing the reader with a clear picture of the considerations and challenges from research and discovery through clinical development and finally to commercialization. Critical points of engagement with the FDA, including approval to initiate first-in-human clinical studies (via Investigational New Drug application or IND) as well as approval to distribute the product across the country (Biological License Application or BLA), are discussed. Thus, this chapter provides an overview of the regulatory requirements for biological product development and approval, together with specific examples that provide a clear understanding of the different stages involved.

CHAPTER 7: COMBINATION PRODUCTS

This chapter provides a comprehensive review of the fundamental definitions and complex regulatory pathways related to combination products. The author provides an expansive summary of the fundamental definitions of combination product categories, together with explanations related to processes and FDA centers that review the IND and marketing applications according to multiple defined sets of criteria, as well as the need for case-by-case reviews on the basis of primary classification of each combination product. Various detailed examples are provided to clarify the different definitions.

The author defines a combination product as the composition of two or more different types of medical products, which can include a combination of a drug, device, and/or biological product with one another. This is different from a drug product containing two or more active pharmaceutical ingredients, which, legally, is not a "combination product" but rather a "drug product which includes a fixed combination of active ingredients."

The chapter provides a section on the classification of combination products to clarify when a combination product is considered a drug, biologic, or device. In addition, this section provides guidance in cases lacking clarity on the classification of the combination product and where there is a need to submit a request for designation (RFD) to the FDA's Office of Combination Products (OCP).

The author further describes the role of the OCP as the key office providing guidance in the regulation of combination products, as well as the main office that assigns a specific center with primary jurisdiction for the review of a combination product.

The chapter describes the type of investigational and market application needed for combination products depending on the constituents of the product. The regulatory pathways will lead to New Drug Applications (NDAs), Biological License Applications (BLAs), Premarket Notifications (510(k)), or Premarket Approval (PMA) applications for medical devices. The responsible center within the FDA will be assigned to review the applications. In general, regulatory review of combination products is a complex task because of the potential for oversight overlap and the increasing complexity of small molecule drugs, biologics, and medical devices. Small molecule drugs are primarily regulated under section 505 of the Food, Drug and Cosmetic (FD&C) Act and reviewed by the Center for Drug Evaluation and Research (CDER), and biologics are regulated under section 351 of the Public Health Services (PHS) Act and reviewed by either CDER or the Center for Biologics Evaluation and Research (CBER). The Center for Devices and Radiological Health (CDRH) regulates certain biological and combination products.

Importantly, this chapter provides a summary of the multiple regulatory requirements for combination products, including manufacturing, human factors, and clinical investigations, as well as post-marketing safety reporting, post-marketing product changes, packaging requirements, and labeling.

Multiple case studies of FDA-cleared products are also summarized for the reader to clearly obtain a comrehensive understanding of the application processes and requirements based on the product definition, classification, regulatory pathway, responsible agency center, and overall recommended activity, e.g., clinical study.

In summary, this chapter provides invaluable information for any entity involved in the development and commercialization of combination products.

CHAPTER 8: FDA-EXPEDITED PATHWAYS

This short chapter provides a practical and comprehensive review of the various FDA programs that provide incentives for expediting product approvals for drugs, biologics, and medical devices. Several pathways exist today to facilitate and accelerate development. These pathways are discussed in detail and include the following designations: Fast Track, Breakthrough, Priority Review, Accelerated Approval Pathway, Orphan Drug, Regenerative Medicine Advanced Therapy (RMAT), Rare Pediatric Disease, Qualified Infectious Disease Product, and Breakthrough Device. A table with an excellent comparison of the different FDA expedited programs is presented, allowing for a quick reference guide to the different options. Thus, this chapter provides a guide to facilitate a straightforward understanding of the different programs the FDA has implemented to expedite product approvals for drugs, biologics, and medical devices, which are key steps to consider from an early stage of product development.

CHAPTER 9: FDA AND INVESTIGATOR-INITIATED SPONSORED RESEARCH

This chapter primarily focuses on investigator-initiated research and the regulatory implications associated with initiating and conducting clinical research. Regulations for investigational drugs or medical devices define the roles of the parties involved, including the principal investigator (PI) and sponsor. Researchers may perform research with a commercial sponsor that may hold an Investigational New Drug (IND) or an Investigational Device Exemption (IDE). When the investigator serves as both the PI and the sponsor, or sponsor-investigator (SI), it is of critical importance to understand those dual roles to prevent any potential non-compliance. Oftentimes, the investigator may not be aware of this dual role as a sponsor-investigator and thus not fully understand the regulatory, financial, contractual, and ethical implications. The researcher needs to be able to define his/her role in order to prevent any misconception or misperception of a conflict of interest that may lead to bias and may even eventually prevent the research from advancing. Therefore, this chapter emphasizes the critical importance of transparency in declaring investigational roles, as those declarations will be eventually scrutinized as part of an FDA application or oversight.

Sponsors and investigators have a number of critical regulatory obligations that need to be followed. These obligations and responsibilities are summarized in the FDA Forms 1571 and 1572 and need to be signed by the sponsor and investigator, respectively. Some of these obligations include study risk determination, IRB approval, safety reporting (serious adverse events (SAEs) and unexpected adverse effects (UAEs)), filing annual reports, and a number of disclosures including financial ones, among other obligations. In the case of devices, the sponsor or sponsor-investigator is responsible for assessing the risk of a medical device study and presenting such risk to the IRB. A medical device study could be deemed non-significant risk (NSR) or significant risk (SR). Regardless of the risk determination, the applicable FDA regulations should be followed for all FDA-regulated drug and medical device studies. Failing to comply can have catastrophic consequences. It is typically recommended that the sponsor-investigator consider including a regulatory specialist or consultant on the team to comply with all the FDA requirements. The role of the researchers and sponsors in a clinical study needs to be documented and classified by responsibilities as they relate to the FDA requirements.

In summary, this chapter is dedicated to understanding the regulatory compliance considerations that need to be taken by a sponsor-investigator considering initiating and conducting a clinical trial. These include financial sponsor considerations, protocol development, monitoring obligations, the affiliated institution's capacity to support clinical investigations (i.e., support staff, equipment, general facility qualities), legal and financial considerations, as well as IRB evaluation requirements.

CHAPTER 10: MEDICAL COUNTERMEASURES FOR PUBLIC HEALTH EMERGENCY PREPAREDNESS AND RESPONSE IN THE UNITED STATES: A REGULATORY PERSPECTIVE

This chapter provides a comprehensive summary of the US regulatory considerations for medical countermeasures (MCMs) for public health emergency preparedness and responses. The chapter introduces a short historical synopsis of the legislative efforts related to preparedness for nationwide public health emergencies (PHE), including those declared by the US Secretary of Health and Human Services (HHS). Special emphasis has been given to regulatory pathways associated with COVID-19 as an unprecedented case study for which multiple novel drug products and diagnostic tools became available to the public through the Emergency Use Authorization (EUA). For instance, on January 31, 2020, under Section 319 of the Public Health

Service (PHS) Act, a PHE was declared by the HHS due to confirmed cases of COVID-19. The success in the coordination and approval of MCMs is a testament to the concerted multiagency efforts that were made over the past two decades to focus on this kind of complicated and extreme situation, including the implementation of the Strategic National Stockpile (SNS). Other notable examples of MCMs are also introduced, including threats due to other viruses, bacteria, fungi, and non-living products that may be of biodefense concern. Such threats are based on risk assessment of the pathogenicity, virulence, lethality, infectivity, and transmissibility of the agents affecting citizens domestically and abroad. Examples of these threats have included viruses such as SARS-CoV-2, Middle East respiratory syndrome coronavirus (MERS), Ebola virus, Zika virus, H7N9 influenza virus, enterovirus D68, and H1N1 and H5N1 influenza viruses. Moreover, other biothreat examples include the anthrax toxin produced by *B. anthracis*, botulinum neurotoxins produced by certain *Clostridium* bacteria, as well as ricin, which naturally occurs in the beans of the castor bean plant, *Ricinus communis*. The chapter is therefore devoted to providing unique knowledge on regulatory considerations for MCMs against biological agents and toxin threats, including vaccines, monoclonal antibodies, antivirals, and antitoxins, such as insights on the FDA regulatory designations and pathways for MCMs. (For a more in-depth focus, the reader is referred to Chapter 8, "FDA-Expedited Pathways," by Oliver V. Cácananta.)

The mission and organization of the Public Health Emergency Countermeasure Enterprise (PHEMCE) are described in detail, including the role played by other federal agencies such as the Centers for Disease Control and Prevention (CDC), National Institutes of Health (NIH), and Department of Defense (DoD). Examples of activities performed by the different agencies involved in PHEMCE are provided, including the role of the NIH in early research during the COVID-19 pandemic, developing COVID-19 diagnostic kits as well as vaccines, monoclonal antibodies, and other COVID-19 MCMs. The role of the Accelerating COVID-19 Therapeutic Intervention and Vaccine partnership, including cross-collaboration among federal agencies, is mentioned as a key element to accelerate the clinical development of COVID-19 vaccines.

The role of the DoD through Defense Advanced Research Projects Agency (DARPA) in the development of breakthrough technologies against COVID-19, such as the Pandemic Prevention Platform, is mentioned as an example of the key role that the DoD plays in developing innovative technologies within the MCM space. Finally, the chapter provides a summary of PHMCE development and manufacturing capabilities. Here, the function of the Biomedical Advanced Research and Development Authority (BARDA) to support the advanced development of critical MCMs is described.

Overall, this chapter provides historical references and examples of the multiagency effort in dealing with biothreats, as well as challenges and future directions for investing in overcoming supply-chain issues and further improving the emergency response.

2 Dancing with the FDA
Managing Expectations

Raymond Vennare

Whether an academic research institution or a university incubator, a multi-billion dollar publicly traded company or an early-stage biotech in need of financing, no segment of the biotechnology community is untouched by the reach, discretion, authority and control of the FDA.

This is an important consideration for many reasons, not the least of which is the imperative to establish relationships, educate novitiates and manage expectations from discovery through commercialization. It is virtually impossible for the FDA, or any oversight agency with such broad regulatory powers over so many diverse disciplines, to know and understand each of these disparate disciplines substantively, equally or well. No agency, university, company or country possesses that much domain expertise in one place, at the same time, to fully comprehend these issues.

It is also true that the FDA simply cannot know as much about any particular discovery or technology as the person, or people, who have discovered or developed it.

For more than two decades, either directly or tangentially, I have been involved in the development of novel technologies, new products and emerging companies spanning the entire spectrum of biotechnology, from informatics and diagnostics to therapeutics and medical devices. As an investor, board director or senior executive, I have come to understand the roles and responsibilities of the FDA and to anticipate the reverberating impact of that agency at every inflection point along the continuum of research, development and clinical adoption. This ripple effect not only touches patients and clinicians but empowers or constrains entrepreneurs, innovators and product developers, especially with respect to the business of biotechnology.

The consequence of that regulatory oversight, whether enabling or inhibiting, is material and far-reaching; it permeates every aspect of discovery, research, patentability, capital formation, company creation, reimbursement and, ultimately, the healthcare system itself. The very nature and quality of medical care afforded to the citizens of this country significantly depend upon approval, clearance or access to technologies and therapies subject to the jurisdiction or discretion of the FDA.

This chapter is intended to inform those researchers, inventors, founders and entrepreneurs who may be unfamiliar with crucial business issues related to regulatory compliance. It serves as a primer for those less experienced with the process of moving a medical or technical breakthrough from the laboratory to the marketplace. It is neither hypothetical nor theoretical; it is applied.

While obliquely touching on some of the larger and more complex issues of regulatory compliance, however, this discussion avoids any debate regarding the

DOI: 10.1201/9781315117911-2

underlying policies, procedures, politics or organizational capacity of the FDA to fully comprehend, assess or manage the voluminous filings, submissions and applications currently in the pipeline in a timely, consistent and reliable way.

Rather, this critique embraces these challenges, accepts them as fact and provides some very basic guidelines for identifying, anticipating and mitigating these issues before they become a problem. It is about understanding rules of engagement and accepting roles and responsibilities. It endorses proactive communication, if not the very real need to nurture and maintain mutually beneficial relationships between the FDA, industry and entrepreneurs.

THE DANCE

Friedrich Nietzsche is quoted as once having said, "those who were seen dancing were thought to be insane by those who could not hear the music." Whether apocryphal or not, what I have come to clearly understand is that interactions between the FDA and industry are indeed a form of choreography – typically difficult, sometimes contentious and often rather tenuous. The precariousness of this relationship is primarily because of the inherent struggle between government oversight and market forces; between the power to monitor or regulate discovery and the need to recoup the cost of funding that discovery; and the obligation to control what products enter the market and the fiduciary responsibility inherent in monetizing those products fairly.

Time, in fact, has emerged as the greatest enemy of progress and foretells the highest risk of failure. The longer it takes to make a decision, the more onerous the outcome. The more constrained the process of licensing intellectual property, manufacturing iterative technologies or launching companies around novel discoveries, the fewer choices and options are available to physicians, patients, clinicians, researchers, inventors and engineers. In other words, the passage of time puts on hold certain definitive actions and decisions by those very people who drive discovery and innovation and, more importantly, by those who might otherwise most benefit from these costly and resource-intensive efforts.

Literally, figuratively and metaphorically, this is both a matter of survival for patients and survivability for companies, for quality of care and quality of life. Anything that one can do to better understand, manage or expedite this process can only improve the intended outcome.

To a certain extent, this dichotomous relationship between the FDA and industry is completely understandable. It is the psychological equivalent of potential energy versus kinetic energy. Potential energy, in its most basic form, refers to the stored capacity of an object to "potentially" change the state of another object; potential energy is a conservative force. Kinetic energy, on the other hand, specifically signifies energy in motion; kinetic energy is catalytic.

The metaphor becomes obvious when you change the word "energy" to "agency" or "industry" respectively. The FDA, as a government agency, possesses enormous potential energy but remains essentially passive and static until some other force is applied. Industry, on the other hand, is kinetic, if not frenetic. It is energized by its own motion and stalls only when that momentum is dampened or interrupted. Industry, by temperament and necessity, is proactive. The entire FDA review process, however, is reflexive and reactive.

A similar case can be made for the entrepreneur, about whom Peter Drucker said, "always searches for change, responds to it and exploits it as an opportunity."

ENTREPRENEUR VS. ENTREPRENEURSHIP

It is sometimes said that successful entrepreneurs are born, not made – that, like Athena sprung fully formed from the head of Zeus, they are born to do battle. Others insist that entrepreneurship is a learned skill – that it can be codified, taught and passed from one generation to the next. It presumes nature over nurture, pits evolution against inheritance, and challenges Charles Darwin to an entrepreneurial duel with Jean-Baptiste Lamarck.

Experience tells me that the answer to this question depends entirely upon how the question is asked. If one asks, "Can entrepreneurship be taught?," the answer is yes. If one asks, "Can you can teach someone to be an entrepreneur?," the answer is no. The distinction is subtle but extremely important. While many people can, in fact, be taught certain skill sets that engender entrepreneurial behavior, not everyone is capable of living the life of an entrepreneur.

In the same way that anyone can be taught the fundamentals of baseball – to understand the rules, run the bases and play various positions – not everyone is a natural-born athlete. I can hit the ball to center field, but that does not make me Babe Ruth.

To *be* an entrepreneur (noun) is very different from *being* entrepreneurial (adjective). The former is a state of being; it is a temperament defined by intuition, determination, vision, intestinal fortitude and a high degree of risk tolerance. The latter is a methodology; it is a set of learned skills that can be reasoned and applied.

In any event, what Drucker is suggesting is that, without the ability to exploit opportunities and to effect the change necessary to realize the full potential of those opportunities, the majority of innovation in this country would slow to a crawl. Research, development, discovery, jobs, companies and new product development would atrophy and wither away. Deal flow would dry up for investors, strategic partners would see a drop in product pipelines, and fewer acquisition, merger and IPO targets would catch the eye of Wall Street.

"Systematic innovation," Drucker said, "consists in the purposeful and organized search for changes and in the systematic analysis of the opportunities such changes might offer for economic or social innovations."

The ability to advance innovation in the service of social good, at least with respect to the business of biotechnology, has everything to do with our topic.

TECHNOLOGY COMMERCIALIZATION

I previously referred to the continuum from discovery through commercialization as an important consideration with respect to the evolution and fruition of ideas.

But mastering the mechanics of this evolutionary process is at least as important as understanding the inevitable interactions with the FDA itself. In point of fact, regulatory compliance, while critical and unavoidable, is only one of many considerations with respect to the commercialization of most, if not all, biomedical products. These products, by definition and designation, are actually quite broad. The FDA guidelines, for example, define a medical device as:

> an instrument, apparatus, implement, machine, contrivance, implant, in vitro reagent, or other similar or related article, including a component part, or accessory which is: recognized in the official National Formulary, or the United States Pharmacopoeia, or any supplement to them, intended for use in the diagnosis of disease or other conditions, or in the cure, mitigation, treatment, or prevention of disease, in man or other animals, or intended to affect the structure or any function of the body of man or other animals, and which does not achieve any of its primary intended purposes through chemical action within or on the body of man or other animals and which is not dependent upon being metabolized for the achievement of any of its primary intended purposes.

In other words, anything that is used on, in or around any living animal or human being may be construed as a medical device. The scrutiny that is applied to the evaluation of these devices falls within three distinct categories: Class I, Class II or Class III. These classifications indicate the presumed level of patient risk associated with that device and the concomitant approvals required to legally market that device.

Class I, for example, suggests minimal potential harm to the user and focuses primarily on design, manufacture and history of use. Class II implies an elevated risk to safety and certainly would require mitigating controls or special conditions; these might include modified labeling, mandatory performance standards or post-market assessments. The most rigorous and severe scrutiny is reserved for Class III medical devices. Generally speaking, this level of classification typically refers to devices that might be viewed as primarily intended to sustain human life or which might otherwise pose an unreasonable risk to patients.

From the perspective of technology commercialization, as one might imagine, these regulatory burdens directly impact the methodology, metrics and milestones associated with bringing these devices to market. They also inform the business strategy, not the least of which include company structure, capital formation, internal resources, financial assumptions, market potential and exit criteria.

Given this backdrop of compliance and commerce, the viability of any technology, the sustainability of all ventures or monetization of most assets will require a thorough analysis of regulatory pathways, reimbursement codes, market drivers, investor expectations and many other issues surrounding intellectual property or freedom to operate. Together, these represent the most fundamental assessments and influential assertions, at least in terms of due diligence, that any strategic partner or stakeholder will expect to see. This is true whether it is a license, a technology or a company. Determining viability, sustainability and marketability, however, is a completely different matter.

Again, in the context of technology commercialization, virtually all discovery moves along a similar path from basic research through translational science – that is, from proof of concept (scientifically defensible) to marketability (commercially viable). The focus here is on validating relevant scientific principles as they relate to a commercial application. Assuming that the science is right and the application is apparent, the question then turns from the viability of the concept to the scalability of the application.

This is the beginning of the commercialization process. By "commercialization," I mean both the methodology by which a scientific principle is transformed into a scalable product and the means by which that transformation takes place. Broadly speaking, this is what separates research from development, academia from industry and experimentation from product development. Technology commercialization, by definition, proclaims the intent to market a product grounded in novel or iterative discovery and protected by intellectual property for the purpose of making money.

The greatest hurdles to overcome with respect to the commercialization of any medical technology are regulatory compliance, reasonable reimbursement, defensible intellectual property and reliable market intelligence. Identifying and mitigating these risks are essential strategic considerations in anticipation of licensing an asset, manufacturing a product or launching a company.

Whether a license, a product or a company, however, the mitigation of these risks will equally determine the likelihood of attracting investors, strategic partners and customers. And, without first having cleared these hurdles and mitigated these risks, I would argue that raising capital, finding partners and solidifying a customer base would be a be difficult, if not impossible, task.

More importantly, perhaps, is that all of these considerations will fundamentally inform the decision-making process: from drafting a business plan to designing a protocol; from defending financial assumptions to identifying likely acquisition targets; and from filling board seats to hiring legal representation.

Whether a start-up laboratory or a publicly traded multinational conglomerate, no one escapes the regulatory process. It is ubiquitous, persistent and predictably unpredictable. For no other reason than practicality, any company (early-stage or otherwise) should follow legislation, anticipate delays, plan for the inevitable cost of ongoing engineering and be prepared for a prolonged regulatory review process. The more highly regulated the class of device – Class I, Class II, or Class III – the more certain the scrutiny.

BACK TO THE FDA

First and foremost, the FDA is a firmly ensconced government agency. It exists to verify, validate and regulate food, drugs, medical devices, radiation-emitting products, vaccines, blood, biologics, cosmetics and tobacco products manufactured (human or animal) in the United States. The two immutable principles by which all products, treatments, therapies, drugs and devices will be judged are safety and efficacy.

More importantly, one should appreciate the fact that the FDA operates under Administrative Law.

That is, while the FDA answers to Congress, it operates independently. It can change regulations, requirements, definitions and guidelines at will and without notice. And this is the single greatest fear of any company of any size, regardless of the product, with respect to developing a sound and reliable regulatory strategy.

Moreover, the ability, capacity or willingness of the FDA to clear or approve a plethora of drugs, therapies and medical devices, in the context of patient outcomes, cannot be underestimated. In effect, the FDA itself has the power to influence, improve or impede the availability and use of various tools, devices and treatment options upon which physicians and patients may eventually rely.

Ironically, the pay-for-performance model inherent in the Affordable Care Act (ACA) is predominantly outcomes-based. If this model is to work properly, then the way in which medical facilities function, data is collected and patient information is shared requires a systemic and sustained transformation with respect to hospital workflow, electronic health records, remote body sensors, wireless drug interaction monitors, device utilization metrics, data-driven care regimens, healthcare IT modifications, bioinformatics applications and telehealth services.

Outcomes-based reimbursement, which replaces volume-based payments, will now be governed by Accountable Care Organizations (ACO), which, in turn, are responsible for coordinating Medicare and Medicaid payment policies. The Center for Medicare and Medicaid Innovation will test the new payment and delivery systems, while the Independent Payment Advisory Board (IPAB) will focus on Medicare. The last of these three governing bodies, the Patient Centered Outcomes Research Institute (PCORI), will evaluate, assess and provide comparative analysis.

While these changes will undoubtedly provide a multitude of opportunities for academia and industry to solve problems, meet needs and commercialize technologies, many of these same products or services will surely be scrutinized, regulated, governed and guarded by the FDA.

If you think about discovery and innovation as a funnel, and the FDA as a filter, you begin to see the problem. New ideas, discoveries, technologies and processes are poured into a gaping regulatory funnel on a daily basis. This represents a massive amount of data that has been collected, curated and analyzed over many years of research by hundreds of individuals who have dedicated their entire professional lives to learning and discovery. Whatever that revelation might be – a gene, a protein, a pathway, a device, a mechanism of action – the details of that information are best known to that person and understood by those people.

As the funnel narrows, a bottleneck is formed. The filter is unable to sift through all of this new data because it lacks the experience, resources, context and expertise necessary to make informed decisions. It is unsure of who should evaluate which technology, or how one discovery, impacts, enables or invalidates another. Weeks pass. Months pass. More information is requested. Arguments ensue and rationales are proposed.

While in the opening paragraphs of this chapter I proposed to avoid this very debate about the policies, procedures, politics or organizational capacity of the FDA,

I do believe that it might be useful to at least provide some context regarding these issues. My goal, as the title of this chapter suggests, is not only to share certain insights but also to manage expectations.

As mentioned earlier, the commercialization of all biomedical technology moves along a prescribed path of research and development regardless of its origin – a university laboratory, a commercial research facility, a medical device manufacturer or an inventor in his or her own garage. First, there is basic research, followed by translational science, leading to technology transfer and, ultimately, product development.

In drafting this narrative, I reached out to several people in academia and industry for their feedback and perspective. I wanted to better understand how my views agreed with, or contradicted, the experiences of other professionals with ties to biomedical research and commercialization, specifically as it related to their personal interactions with the FDA.

Everyone with whom I spoke related similar experiences, rendered analogous opinions and offered comparable recommendations. Words like "unpredictable," "obsolete," "overwhelmed" and "impractical" were frequently used in conversation when referring to the FDA. These are words, by the way, that no researcher, investor, CEO or board member ever wants to hear when planning, funding, commercializing or launching any medical technology.

Nevertheless, there is a clear mandate that the primary obligation of the FDA is to protect the safety and ensure the efficacy of any product or device marketed to the general public. On the other hand, there is a deep sense of frustration and anxiety that no comparable requirement seems to exist in deference to the legions of innovators and developers of new technologies in terms of reasonable, realistic and consistent regulations by which those novel discoveries might be assessed, cleared or approved.

This is made most apparent as new strategies are emerging for the treatment and care of patients, where combination products have shown promise. Combination products, as defined by the FDA, are "therapeutic and diagnostic products that combine drugs, devices, and/or biological products."

By their own admission, the FDA expects to receive,

> large numbers of combination products for review as technological advances continue to merge product types and blur the historical lines of separation between FDA's medical product centers, which are made up of the Center for Biologics Evaluation and Research (CBER), the Center for Drug Evaluation and Research (CDER), and the Center for Devices and Radiological Health (CDRH). Because combination products involve components that would normally be regulated under different types of regulatory authorities, and frequently by different FDA Centers, they raise challenging regulatory, policy, and review management challenges. Differences in regulatory pathways for each component can impact the regulatory processes for all aspects of product development and management, including preclinical testing, clinical investigation, marketing applications, manufacturing and quality control, adverse event reporting, promotion and advertising, and post-approval modifications.

Simply translated, this means that the FDA itself has yet to determine who will review these product applications, how they will be evaluated or how long that might take.

From the perspective of an entrepreneur, investor or strategic partner, this ambiguity makes it virtually impossible to reliably anticipate timelines, milestones and funding requirements in the development of protocols or product development plans. This means that, whatever promise these combination products might have with respect to the treatment and care of patients, the likelihood of advancing these products into the marketplace will be long, arduous and costly. For larger firms flush with cash, this is less of an issue, but for early-stage companies trying to raise capital, it is almost impossible.

Conversely, companies that have already raised capital and are actively committed to the commercialization of combination products face the very real possibility that FDA approval will be stalled, guidelines changed or rules applied that might literally put them out of business.

MANAGING EXPECTATIONS

It is interesting to compare the current frustrations of a first-time entrepreneur in an early-stage biomedical company with the historical perspective of more seasoned professionals in the medical device industry.

Michele Marcolongo, PhD, is the Drosdick Endowed Dean of the College of Engineering and former Senior Associate Vice Provost for Translational Research and Professor of Materials Science and Engineering at Drexel University. Dr. Marcolongo is also an engineer, researcher, inventor and entrepreneur. In that capacity, she personifies the entire continuum from basic research to potential commercialization and, with that perspective, demonstrates the need to have an open dialogue with the FDA about an appropriate and defensible regulatory strategy, evaluation process and study design. From her perspective, the earlier in the process that this occurs, the better researchers and companies can prepare. Equally important is that some assurance is given that, once those standards are established, those criteria remain unchanged with respect to the chosen regulatory pathway, product development plan, financial assumptions and funding milestones. Most crucially, however, is that those criteria and standards are fairly and consistently applied across all other (potentially competitive) technologies.

This neither challenges, changes nor invalidates the mission of the FDA but rather removes the risk of uncertainty in the development process, a process which fundamentally allows the research team to reliably streamline an efficient course of action, lower risk and attract capital. Moreover, to the extent that these regulatory changes may restore investor confidence and presumably expedite market approval, they may also help expand the quality and reach of healthcare in this country.

Paul Citron, PhD, founding member of the American Institute for Medical and Biological Engineering and member of the National Academy of Engineering, not only shares many of the views already expressed in this chapter but informs the way in which they have been articulated. Dr. Citron, now retired after a 32-year career at Medtronic, Inc., sees a much deeper and more systemic problem: that the FDA simply lacks the necessary internal technology-specific expertise to evaluate novel and innovative discoveries and cannot reasonably be expected to acquire it.

Politics aside, the FDA's sphere of influence, administrative powers and broad control (25% of the US GDP) is simply too great to deal with such an expansive portfolio of regulated products, products which include food, tobacco, drugs, medical devices, veterinary products, biological and cosmetics.

The end result, Citron concludes, is that the approval process becomes so unpredictable, overly burdensome and slow that it causes US patients to be among the last in the world to gain access to life-saving technologies. In some instances, developers are electing to bypass market entry in the US for innovations funded by federal grants because of the regulatory situation here.

More damaging is that delayed patient access, often measured in years compared to Europe and elsewhere, is hurting patients.

Any one of these topics is a dissertation in itself, but I will end this article as I began: to reiterate that the role and reach of the FDA touches every inflection point along the continuum of discovery through commercialization – not only on patients and clinicians but with respect to entrepreneurship, innovation and product development, especially as it relates to the business of biotechnology. That reach touches every aspect of discovery, research, patentability, capital formation, company creation, reimbursement and, ultimately, the healthcare system itself.

To the extent that this chapter is intended for those researchers, inventors, founders and entrepreneurs who may be unfamiliar with crucial business issues related to regulatory compliance, it may also serve as a primer for those less experienced with the process of moving a medical or technical breakthrough from the laboratory to the marketplace.

It's about understanding the rules of engagement and accepting those roles and responsibilities. It's about proactive communication and the very real need to nurture and maintain mutually beneficial relationships. It's about knowing when to follow and learning how to lead.

It's about dancing with the FDA and not stepping on their toes.

3 New Drug Application

Harvey M. Arbit and Wrenda Teeple

INTRODUCTION

The FDA is a federal consumer protection agency that was created in 1906 with the passage of The Pure Food and Drug Act. That federal law prohibited interstate commerce (i.e., manufacture, sale, or transportation) of adulterated or misbranded foods or drugs; it required that the presence and amount of selected dangerous or addictive substances, such as alcohol, morphine, heroin, cannabis indica, chloroform, and cocaine, be labeled; and it identified the United States Pharmacopoeia (USP) and the National Formulary (NF) as official standards for drugs.

In September 1937, S.E. Massengill Company in Bristol, Tennessee, introduced Elixir Sulfanilamide. This was a new oral dosage form of an already marketed anti-infective drug, previously only available as a powder and tablet. The firm used an untested solvent, diethylene glycol, and raspberry flavoring to improve the drug's palatability. By the time diethylene glycol was identified as the poisonous component and the FDA was notified, the product had caused 107 deaths, including many children. This led to Harold Watkins, the chemist at S.E. Massengill who developed the elixir, taking his own life less than two years after developing the elixir. At that time, the firm could only be prosecuted for distributing a misbranded drug. An "elixir" had to contain ethanol as a solvent. This tragedy led to the passing of the Food, Drug, and Cosmetic Act (FDCA) on June 25, 1938. This act required that firms prove to the FDA that any new drug was safe before it could be marketed. This was the beginning of the New Drug Application (NDA). In addition, the new law covered cosmetics and medical devices, authorized factory inspections, and outlawed unfounded therapeutic claims for drugs. The FDA began enforcing the FDCA in a manner that effectively limited direct-to-consumer availability of some drugs based on the decision by the FDA that some drugs could *not* be labeled safely for consumers, despite the law not explicitly giving that authority to the FDA. In order to clarify the role of the FDA in determining and regulating physician-ordered (Rx) and over-the-counter (OTC) drugs, the Durham-Humphrey Amendment passed in 1951. This amendment to the FDCA identified clear parameters for what constitutes a prescription drug, who would be responsible for identifying such drugs, and the conditions under which a prescription could be refilled. This removed freedoms from the consumer to comprehend accurate drug labeling and placed sole authority on appropriate prescription drug use on the physician.

As drug developers, manufacturers, and prescription drug use came under more scrutiny, additional unregulated issues in the pharmaceutical industry were

 DOI: 10.1201/9781315117911-3

identified, including drug effectiveness and drug advertising. On September 12, 1960, an American licensee, the William S. Merrill Company of Cincinnati, submitted a new drug application for Kevadon, the name brand of a sedative thalidomide, which had been marketed in Europe since 1956 as a sleep aid and anti-emetic. The anti-emetic uses were significant for pregnant women due to its efficacy in treating "morning sickness" and its over-the-counter availability in much of the world. The FDA medical officer in charge of this review, Dr. Frances Oldham Kelsey, believed that the safety data supporting the use of this drug was incomplete and refused approval without additional safety studies. In November 1961, the drug was pulled from the German market because of its association with grave congenital abnormalities. Several thousand newborns suffered the drug's teratogenic effects. In the United States, Merrill Company and Smith, Kline, and French (SKF) had distributed this drug to over 1,000 physicians under the purview of investigational use. Over 20,000 Americans received thalidomide from these companies or obtained the drug overseas, including 624 pregnant patients. About 17 known newborns in the United States suffered the teratogenic effects of thalidomide, including facial palsies, congenital heart disease, internal organ damage, and disfigurements such as phocomelia, a rare congenital condition that causes absent or severely underdeveloped limbs. This tragedy led to the Kefauver-Harris Amendments, which became law on October 10, 1962. This law required that all manufacturers had to prove to the FDA that their drugs were effective as well as safe before they could be marketed, including all drugs introduced between 1938 and 1962 now having to now show efficacy. This law also provided the FDA authority to regulate advertising of prescription drugs and control clinical investigations, including the requirement for informed consent. In addition, the FDA received authority to establish good manufacturing practices (GMP) to promote quality assurance and access company control and manufacturing records to verify production procedures.

As we entered the 1990s, the current standard for drug testing and approval evolved to include pre-clinical research and development, such as animal testing. Once a firm files an investigational new drug (IND) application with the FDA, Phase 1 clinical trials can begin to study the drug's toxicity and pharmacology (i.e., safety) in humans. If Phase 1 studies support the drug's safety, Phase 2 clinical trials may proceed to demonstrate efficacy in the target patient population at the dosing regimens studied for safety. Once safety and efficacy have been demonstrated in Phases 1 and 2 clinical trials, the drug may proceed to Phase 3 large pivotal studies required for final review and evaluation by the FDA. After the new drug application (NDA) has been approved, the drug can be lawfully marketed for the indications as described in the labeling.

In 1984, the Drug Price Competition and Patent Term Restoration Act (Hatch-Waxman Act) was enacted to expedite the availability of less costly generic drugs by permitting the FDA to approve generic versions of brand-name drugs without having to repeat the research to demonstrate safety and efficacy.

In the 1980s and 1990s, several other significant changes took place in the drug approval process, such as treatment INDs (1987), accelerated approval mechanisms (1988–1992), and parallel track investigations (1992). This dramatically increased the submission of new drug applications to the FDA, subsequently leading to the Prescription Drug User Fee Act of 1992 (PDUFA). This allows the FDA to charge fees to sponsors to support additional resources for the drug review process, eliminating the backlog of pending applications and decreasing the time for FDA review and approval of new drug applications.

The FDA's website contains a vast amount of information that can be useful when developing new drugs. Each calendar year, the Center for Drug Evaluation and Research (CDER) publishes a table of "Drug and Biologic Calendar Year Approvals," which lists all NDA approvals for the calendar year and clearly identifies which drugs are New Molecular Entities (NMEs) and which drugs have been granted priority review and/or orphan designation (USFDA 2020). The FDA also makes approved NDAs publicly available in a database that includes the NDA number, type of submission, NDA classification, approval date(s) and history, letters, reviews, labels, priority review status, and/or orphan drug status (USFDA 2021b). Initial NDA review documents include product quality review, multi-discipline review (consisting of summary review, office director review, cross-discipline team leader review, clinical review, non-clinical review, statistical, and clinical pharmacology review), proprietary name review, officer/employee list, other reviews, risk assessment and risk mitigation review, and administrative and correspondence documents. These documents are redacted so that confidential and proprietary information is not divulged. Since no two drugs are exactly alike, it is always advantageous to communicate with the FDA early in the drug development process. However, you can obtain a high-level understanding of the types of studies that will likely be required for different types of submissions and drugs in various therapeutic classes by reviewing NDAs for similar products.

TYPES OF NEW DRUG ENTITIES

One of the objectives of new drug development is to conduct the right studies the right way at the right times. Another objective is to demonstrate the safe and effective use of the new drug in a cost-effective manner in as short a period of time as possible. To help in the accomplishment of these goals, it is imperative that the new drug sponsor establish dialogue with the FDA early in the development process. The FDA receives more than 100 NDAs in a calendar year (USFDA 2021a). The earlier an item of concern is identified and resolved, the less time it will take for the FDA to complete its review. There was a saying: "For every day the drug is still in FDA review, that is a million dollars in lost sales." It is still true today, only the dollars are larger. One of the most important determinations early in the drug development process is the classification of a new drug product in development. The FDA uses a classification system to categorize drug products. Understanding this system is helpful to prepare for the review process and prioritization of the NDA application. There are ten classifications for new drug applications, described herein.

Type 1 – New Molecular Entity

> A Type 1 NDA is for a drug product that contains a New Molecular Entity (NME). An NME is an active ingredient that contains no active moiety that has been previously approved by the Agency in an application submitted under section 505 of the Act or has been previously marketed as a drug in the United States. A pure enantiomer or a racemic mixture is an NME only when neither has been previously approved or marketed. (USFDA 2015)

An active moiety is the part of a molecule – excluding appended inactive portions – that is responsible for the pharmacological action of a drug substance. Certain drugs are classified as NMEs for purposes of FDA review. Many of these products contain active moieties that have not been approved by the FDA previously, either as a single-ingredient drug or as part of a combination product. These products frequently provide important new therapies for patients. Some drugs are characterized as NMEs for administrative purposes but nonetheless contain active moieties that are closely related to active moieties in products that have previously been approved by the FDA (USFDA 2021c).

Type 2 – New Active Ingredient

> A Type 2 NDA is for a drug product that contains a new active ingredient, but not an NME. A new active ingredient includes those products whose active moiety has been previously approved or marketed in the United States, but whose particular ester, salt, or noncovalent derivative of the unmodified parent molecule has not been approved by the Agency or marketed in the United States, either alone, or as part of a combination product. (USFDA 2015)

Two drug molecules with the same core compound may have different compounds appended to them by either covalent or noncovalent bonds. For example, replacing a hydrogen atom in an acid molecule with *a metal or its equivalent* forms a salt, while replacing the hydrogen atom with *an organic radical* forms an ester. These derivatives may or may not vary from each other in clinically significant ways. This is important as the FDA must assess the novelty of the active ingredient in a new drug, comparing it to a previously approved drug's active ingredient to determine whether the new drug qualifies for the five-year "new chemical entity" (NCE) exclusivity. A drug that contains a single enantiomer of a product that has previously been approved by the FDA for the racemic mixture would be considered a Type 2 NDA.

Type 3 – New Dosage Form

> A Type 3 NDA is for a new dosage form of an active ingredient that has been approved or marketed in the United States by the same or another applicant but in a different dosage form. The indication for the drug product does not need to be the same as that of the already marketed drug product. (USFDA 2015)

The content of the NDA submission for a new dosage form will vary depending on what current data exists for other dosage forms, the similarities and differences between other available dosage forms and the investigational one, and differences in indications. Some dosage forms that are administered orally are considered different dosage forms. For example, solid oral dosage forms may be immediate-release tablets, extended-release tablets, or delayed-release tablets, which are all different dosage forms. There are also different oral formulations (e.g., suspensions, solutions, elixirs, capsules, and tablets) and different routes of administration (e.g., oral, rectal, subcutaneous, intravenous, intramuscular, and topical) that would require a new drug application for a new dosage form. If a sponsor wishes to gain approval for a new route of administration for an already approved drug product and dosage form, the sponsor would submit an efficacy supplement to their approved NDA, per 21 CFR 314.3. Once a new dosage form has been approved for an active ingredient, subsequent applications for the same dosage form and active ingredient should be classified as Type 5 (USFDA 2015).

Type 4 – New Combination or New Ratio of a Multi-Compound Drug

> A Type 4 NDA is for a new drug-drug combination of two or more active ingredients. An application for a new drug-drug combination product may have more than one classification code if at least one component of the combination is an NME or a new active ingredient. The new product may be a physical or chemical combination of two or more active moieties. (USFDA 2015)

In 21 CFR 300.50 (the combination rule), it states that each component must make a contribution to the claimed effects and the dosage of each component (i.e., amount, frequency, duration) in the combination is safe and effective for the intended population. There are numerous examples of combination drug products. Hydrochlorothiazide (HCTZ), a diuretic, has been used to treat hypertension for many decades. It is available in 25 mg and 50 mg tablets. Captopril, an angiotensin-converting enzyme (ACE) inhibitor, has also been used to treat hypertension for many decades. It is available in 12.5 mg, 25 mg, 50 mg, and 100 mg tablets. Often these two drugs are used together to control hypertension. For patient convenience, these two drugs have been combined into one tablet available in four ratios (captopril/HCTZ): 25/15, 25/25, 50/25, and 50/15.

Another example of a Type 4 NDA is Symbyax® by Eli Lilly. It is a combination of fluoxetine and olanzapine. It is available in six strength combinations. The NDA 021520 was approved on December 24, 2003, for the treatment of depressive episodes associated with bipolar disorder. The purpose of evaluating a combination product in a new drug application is to ensure that the appropriate studies, from preclinical safety pharmacology to Phase 3 pivotal trials, are conducted for the proposed combination of drugs and doses. There may be specific concerns with the combination of products that require additional studies which were not needed for the products individually.

Type 5 – New Formulation or Other Differences

A Type 5 NDA is for a product, other than a new dosage form (Type 3), that differs from a product already approved or marketed in the United States in one of the following ways:

1. There is a change in inactive ingredients *that requires either bioequivalence studies or clinical studies for approval.*
2. Another product with the same active ingredient, dosage form, indication, or combination *that requires bioequivalence testing or safety and effectiveness testing for approval.*
3. The product contains an active ingredient or active moiety that has only been approved or marketed as a combination product.
4. The product is a combination product in which one or more active ingredients of an already approved or marketed product is removed, or there is a substitution of a new ester, salt, or other noncovalent derivative of an active ingredient.
5. The product contains a different strength of one or more active ingredients in a previously approved or marketed combination.
6. The bioavailability of the product is different, making the submission ineligible for an ANDA.
7. The product involves a new plastic container that requires additional safety studies. (USFDA 2015)

NDAs may be classified by more than one type. For example, a Type 5 NDA for a combination product in which one or more active ingredients is a new ester, salt, or noncovalent derivative would also be a Type 2 NDA.

An example of a Type 5 NDA is paclitaxel (Abraxane® by Abraxis Bioscience). Paclitaxel (Taxol®) was approved in 1992 as a Type 1 NDA for the treatment of ovarian, lung, and breast cancer. In a Phase 3 randomized study, the mean apparent volume of distribution at steady state, with the 24-hour infusion of Taxol, ranged from 227 to 688 L/m^2, indicating extensive extravascular distribution and/or tissue binding of paclitaxel (BMSC 2011). This extensive distribution and tissue binding led to anaphylaxis and severe hypersensitivity reactions characterized by dyspnea and hypotension requiring treatment. Fatal reactions have occurred in patients despite premedication (BMSC 2011). In order to reduce the occurrence and severity of adverse events, Abraxis Bioscience developed a new formulation of paclitaxel in which the paclitaxel administration to patients with solid tumors is evenly distributed into blood cells and plasma and is highly bound to plasma proteins (94%). This is accomplished by binding paclitaxel to albumin (Abraxis BioScience, LLC 2020). The regulatory strategy here was to submit a 505(b)(2) NDA. This type of NDA allows the application sponsor to reference published literature and the FDA's findings of safety and effectiveness for other products. In other words, the safety and effectiveness of the listed drug do not have to be repeated. It is important to note that a 505(b)(2) application is an NDA for which one or more of the investigations described in section 505(b)(1)(A) of the FD&C Act and relied upon by the applicant for approval "were not conducted by or for the applicant and for which the applicant has not obtained a right of reference or use from the person by or for whom the investigations were conducted" (section 505(b)(2) of the FD&C Act). (https://www.fda.gov/media/100490/download) (Abbreviated New Drug Applications and 505(b)(2) Applications Docket No. FDA-2011-N-0830 September 2016) (https://www.govinfo.gov/content/pkg/USCODE-2023-title21/html/USCODE-2023-title21-chap9-subchapV-partA-sec355.htm).

What is left for the sponsor of the 505(b)(2) NDA are studies necessary to support the change or modification from the listed drug, as well as safety and effectiveness of the new formulation. Complete studies of safety and effectiveness may not be necessary if appropriate bridging studies are found to provide an adequate basis for reliance upon FDA's finding of safety and effectiveness of the listed drug. If this option is being considered for a new formulation of a drug product, early communication with the FDA is recommended to confirm the types of studies required based on current data and to clarify study design expectations to support a new drug application.

Type 6 – New Indication or Claim, Same Applicant

This NDA classification code is no longer used and is replaced with Type 9 and Type 10.

> A Type 6 NDA was used for an NDA received prior to July 27, 2009, for a drug product that duplicates a drug product already approved or marketed in the United States by the same applicant, except that it is intended for a new indication or claim (same active moiety or combination of active moieties, same salt(s), ester(s), or other noncovalent derivative(s), same dosage form, and same formulation (including all ingredients used in the manufacturing process whether or not they are present in the final dosage form). (USFDA 2015)

Type 7 – Previously Marketed but Without an Approved NDA

> A Type 7 NDA is for a drug product that contains an active moiety that has not been previously approved in an application, but has been marketed in the United States. This classification applies only to the first NDA approved for a drug product containing this (these) active moiety(ies). (USFDA 2015)

An example of a Type 7 NDA is for the drug Lanoxin® (digoxin). It is a cardiac glycoside derived from the foxglove plant. Lanoxin brand digoxin was first manufactured and marketed in 1934 by Burroughs Wellcome. In January 1974, the FDA issued a regulation establishing conditions for marketing digoxin products for oral use. The regulation declared all digoxin products for oral use (tablets and elixir) to be new drugs and required submission of ANDAs and bioavailability tests for all oral digoxin products. In September 1993, Glaxo Wellcome (then Burroughs Wellcome) submitted to the agency an NDA (NDA 20-405) for Lanoxin (digoxin) Tablets under section 505(b) of the act. The submission included safety and effectiveness data on the drug product. Based on its review of NDA 20-405 for Lanoxin Tablets, FDA concluded that the application was approvable. This type of new drug application is used less often since many drugs marketed before the New Drug Application process have already submitted the appropriate application and/or documentation to FDA to continue marketing their product. With the passage of the CARES Act on March 27, 2020, several marketed over-the-counter products that have not been formally reviewed by FDA, are not generally recognized as safe and effective (GRASE), and do not have an OTC monograph will require the submission of a new drug application to FDA (Pub. L. No. 116–136).

Type 8 – Prescription to Over-the-Counter (Rx To OTC) Switch

> A Type 8 NDA is for a drug product intended for over-the-counter (OTC) marketing that contains an active ingredient that has been approved previously or marketed in the United States only for dispensing by prescription (OTC switch). (USFDA 2015)

The data required to support an over-the-counter (OTC) drug approval differs from that of a prescription drug because it must be able to be safely used by the general public when used according to the labeling since a prescriber is not involved in ensuring the appropriate use of the product. Many drugs receive prescription use only (Rx only) approval first and submit an application for an Rx to OTC switch with additional post-market data. Type 8 NDAs may be a different dosing regimen, strength, dosage form, or different indication from the previously approved prescription drug product. If there are no differences in the indications, uses, and strengths of a prescription dosage form, the Rx to OTC switch would result in no prescription-only products. In this situation, the application should be submitted as a supplement to the originally approved application. An example of an Rx to OTC switch is Motrin® tablets. In September 1976, McNeil Consumer Products received FDA approval of their Type 1 NDA for Motrin® (ibuprofen) tablets, 400 mg, 600 mg, and 800 mg. The product was available only by prescription. In May 1984, a Type 8 NDA supplement for ibuprofen 200 mg tablets was approved for OTC use.

Type 9 – New Indication or Claim, Drug Not to Be Marketed under Type 9 NDA after Approval

> A Type 9 NDA is for a new indication or claim for a drug product that is currently being reviewed under a different NDA (the "parent NDA"), and the applicant does not intend to market this drug product under the Type 9 NDA after approval. Generally, a Type 9 NDA is submitted as a separate NDA so as to be in compliance with the guidance for industry on Submitting Separate Marketing Applications and Clinical Data for Purposes of Assessing User Fees. (USFDA 2015)

When there is a pending NDA and a Type 9 NDA submission for the same drug product, but different indications or claims, the Type 9 NDA submission will be classified by the FDA the same way the pending NDA is. When one of the two (or more) applications is approved, the unapproved one(s) will be reclassified as Type 9, and then the FDA will "administratively close" the Type 9 application once it is approved. After this occurs, the FDA will only accept submissions to the "parent" NDA or the one approved first.

Type 10 – New Indication or Claim, Drug to Be Marketed under Type 10 NDA after Approval

> A Type 10 NDA is for a drug product that is a duplicate of a drug product that is the subject of either a pending or approved NDA, and the applicant intends to market the drug product under this separate Type 10 NDA after approval. A Type 10 NDA is normally for a drug product that has a new indication or claim, and it may have labeling and/or a proprietary name that is distinct from that of the original NDA. (USFDA 2015)

Similar to the Type 9 NDA, if there is a pending NDA when a Type 10 is submitted, the Type 10 application will be classified the same as the pending NDA. Once one is approved, the other will be reclassified as a Type 10 NDA.

PRECLINICAL SAFETY PHARMACOLOGY AND TOXICOLOGY

Preclinical development of a new drug is used to determine if a product is worth pursuing clinical trials in humans. The goal for drug developers is to identify any potential safety or efficacy concerns that would warrant terminating development. As with most preclinical and clinical study requirements, safety pharmacology requirements have advanced over time. Prior to the 1990s, it was thought that the only drugs capable of inducing the life-threatening cardiac syndrome of torsades de pointes (TdP) were cardiovascular compounds. However, in the mid-1990s, an antihistamine called terfenadine started receiving reports of TdP in the national spontaneous reporting system (SRS). This was a rare condition, only receiving reports of the adverse events after millions of prescriptions had been written. In response to this event and in parallel with the development that specific high throughput screening (HTS) for TdP could identify TdP risk prior to clinical trials, the addition of safety pharmacology to already existing preclinical toxicology became standard in the drug development process. The task at hand for drug developers with safety pharmacology and preclinical toxicology is to present data that supports an unlikelihood that a drug is unsafe, rather than a likelihood that a drug *is* safe. Regulators in the United States, the European Union, and Japan created a working group called the International Conference on Harmonisation of Technical Requirements for Registration of Pharmaceuticals for Human Use (ICH), which develops international standards for industry during drug development to support the use of therapeutics in humans. The term *safety pharmacology* was introduced in ICH *M3 Timing of Nonclinical Safety Studies for the Conduct of Human Clinical Trials for Pharmaceuticals* and *S6 Preclinical Safety Evaluation of Biotechnology-Derived Pharmaceuticals*. ICH has published several guideline documents that provide information on safety pharmacology and toxicology studies such as carcinogenicity, immunotoxicology, pharmacology, biodistribution, genotoxicity, reproductive toxicology, and other toxicity studies.

Non-clinical pharmacology studies can be divided into three categories: primary pharmacodynamic, secondary pharmacodynamic, and safety pharmacology. Safety pharmacology studies should investigate the potential negative physiological effects in the anticipated range of exposure of a substance. Before a substance is studied in humans, it must show sufficient safety pharmacology to support any anticipated dose regimen studied in humans. Carefully designing safety pharmacology studies in parallel with clinical trial design is essential to avoid repeating safety pharmacology studies (USFDA 2011). The kind, duration, and scope of *in vitro*, *ex vivo*, animal, and other tests required to support the conclusion that a substance is reasonably safe to be used in a clinical investigation vary with the nature of the proposed clinical investigations. Which studies to conduct should be selected considering multiple aspects of the drug in development, including the therapeutic class and/or mechanism of action, adverse effects of compounds in the same class, data from enzyme assay or ligand binding studies with a potential for concern, and previous study results from pharmacology, toxicology, or human trials.

Following FDA recommendations on the appropriate design and execution of preclinical safety studies will minimize the unnecessary use of animals and other resources. Sponsors are encouraged to utilize new technologies available that may reduce waste and optimize the use of resources without sacrificing safety or scientific integrity.

In early drug development, there is not always sufficient information to develop a safety pharmacology program based on published literature or previous experience with a drug or drug class. In these situations, it is reasonable for sponsors to take a more general approach to safety pharmacology. Core battery tests are used to study the effects of a substance on vital organ functions for humans – central nervous system, cardiovascular, and respiratory. Other organ systems, such as gastrointestinal and renal systems, should be considered when applicable to a specific drug product or patient population. Consideration should be made regarding *in vivo* animal studies with respect to sample size, use of controls, route of administration, and exposure to the parent compound and metabolites anticipated in humans. The goal of these studies is to determine the dose-response relationship of an anticipated effect at the relative projected exposure levels in humans, given pharmacodynamic differences between species. The type of test, animal model, and study design should be selected in a manner that will provide valuable data for human studies. The FDA follows the Animal Efficacy Rule, which permits the FDA to rely on animal evidence when (1) the agent's mechanism of toxicity is well understood; (2) the endpoints in the animal trials are clearly related to benefit in humans; (3) the drug's effect is demonstrated in a species expected to react similarly to humans; and (4) data allow selection of an effective human dose (USFDA 2021d).

Toxicology studies should be conducted in a manner that supports the proposed use of the drug in humans. This includes acute, subacute, and chronic toxicity tests; tests of the drug's effects on reproduction and the developing fetus (i.e., teratogenicity); any special toxicity tests related to the drug's particular mode of administration or conditions of use (e.g., inhalation, dermal, or ocular toxicology); and any *in vitro* studies intended to evaluate drug toxicity, as applicable.

In vitro studies should be designed to establish a concentration-effect relationship with the desired therapeutic effects and potential toxic effects. The range of concentrations used should be selected to increase the likelihood of detecting an effect on the test system. The upper limit of this range will depend on both the anticipated human exposure range and the appropriateness of that dose in the model, test system, or other assay-specific factors. If there is no effect detected with the highest concentration tested, the decision to discontinue dose escalation should be justified.

In vivo safety pharmacology study data is used to define the dose-response relationship of the adverse effect observed. The onset and duration of the adverse effect should be determined and compared to the dose or concentrations needed to obtain the primary therapeutic effect in the test species, or humans, when possible. The applicability of the pharmacodynamic effects of the drug in the animal model to a human subject should be considered when determining the most appropriate species, dose regimen, and study design. Dose ranges used for *in vivo* safety pharmacology studies may exceed the anticipated therapeutic range in humans to intentionally elicit moderate adverse effects, including dose-limiting pharmacodynamic effects or other toxicity. In the situation that an adverse effect on safety pharmacology endpoints in the test species is not observed with the original proposed dose range, testing of a single group at a limiting dose may be sufficient.

Sponsors should submit justification for the specific studies conducted in IND and marketing applications to the FDA. The FDA will consider waiving certain test requirements if scientific justification can be provided. All safety pharmacology and toxicology studies submitted to the FDA must include the identification and qualifications of the individuals who evaluated the results of such studies, conclude that it is reasonably safe to begin the proposed investigations, and provide a statement of where the investigations were conducted and where the records are available for inspection. Preclinical studies to support an IND or marketing application should be conducted under Good Laboratory Practice (GLP) conditions, or a statement of the reason for noncompliance should be provided to the FDA. GLP regulations are found in 21 CFR Part 58.1 Good Laboratory Practice for Nonclinical Laboratory Studies, which sets a minimum requirement for study conduct, personnel, facilities, equipment, protocols, operating procedures, and reporting for FDA-regulated products.

CHEMISTRY, MANUFACTURING, AND CONTROLS (CMC)

The chemistry, manufacturing, and controls section of the NDA describes the complete manufacturing, packaging, testing, and labeling processes for the drug substance and drug product. The drug substance is also referred to as the active pharmaceutical ingredient (API) and is "intended to have a pharmacological effect or other effect in the diagnosis, cure, mitigation, treatment or prevention of disease, or to affect the structure or any function of the body of man or other animal" (OFR 2019b). This includes compounds that undergo a chemical change in the manufacturing process and are present in the final drug product as the active ingredient. The drug product is the final dosage form that delivers the active ingredient to the study subjects and eventually to patients (i.e., tablet, capsule, patch, suspension, cream, or ophthalmic drops). The drug product may undergo several changes throughout development, some of which may take place during clinical trials. Changes made to the manufacturing process or formulation that impact the drug product should be clearly documented with a justification for the change and submitted to the FDA. The term drug product also refers to a dosage form with no active ingredient but containing excipients and intended for use as a placebo in clinical trials. For example, acetaminophen is the drug substance in the drug product, Tylenol® (acetaminophen) 500 mg tablets.

It is important to keep in mind that the FDA will conduct an establishment or facility inspection prior to NDA approval to ensure that the processes described in the CMC section are the processes that are actually taking place at the facility. Sponsors should be prepared for an inspection of the facility in which the marketed drug supply is manufactured. The FDA offers a number of guidance documents on CMC and GMP for applicant sponsors and drug manufacturers to reference.

The International Council for Harmonisation (ICH) has published many guidance documents that provide information on maintaining standards of quality. These guidance documents include impurities, stability, pharmaceutical development, pharmaceutical quality system, analytical procedure development, lifecycle management, quality risk management, pharmacopeias, and analytical validation (ICH 2021).

The Kefauver-Harris Amendments of 1962 gave the FDA authority to establish good manufacturing practices as a means to promote quality assurance, and access certain company control and production records to verify production procedures.

> CGMP refers to the Current Good Manufacturing Practice regulations enforced by the FDA. CGMPs provide for systems that assure proper design, monitoring, and control of manufacturing processes and facilities. Adherence to the CGMP regulations assures the identity, strength, quality, and purity of drug products by requiring that manufacturers of medications adequately control manufacturing operations. This includes establishing strong quality management systems, obtaining appropriate quality raw materials, establishing robust operating procedures, detecting and investigating product quality deviations, and maintaining reliable testing laboratories. This formal system of controls at a pharmaceutical company, if adequately put into practice, helps to prevent instances of contamination, mix-ups, deviations, failures, and errors. This assures that drug products meet their quality standards. (USFDA 2021e)

The regulation at 21 CFR 314.50 requires that the NDA contain a section describing the composition, manufacture, and specification of the drug substance and drug product. Additionally, guidance documents at fda.gov provide the FDA's current thinking on this topic. The following information should be included:

> A full description of the drug substance including its physical and chemical characteristics and stability; the name and address of its manufacturer; the method of synthesis (or isolation) and purification of the drug substance; the process controls used during

> manufacture and packaging; and the specifications necessary to ensure the identity, strength, quality, and purity of the drug substance and the bioavailability of the drug products made from the substance, including, for example, tests, analytical procedures, and acceptance criteria relating to stability, sterility, particle size, and crystalline form. Similar to the required data for the drug substance, the NDA must contain a list of all components used in the manufacture of the drug product (regardless of whether they appear in the drug product); a statement of the composition of the drug product; the specifications for each component; the name and address of each manufacturer of the drug product; a description of the manufacturing and packaging procedures and in-process controls for the drug product; the specifications necessary to ensure the identity, strength, quality, purity, potency, and bioavailability of the drug product, including, for example, tests, analytical procedures, and acceptance criteria relating to sterility, dissolution rate, container closure systems; and stability data with proposed expiration dating should be submitted for the drug product.

The NDA applicant may provide alternatives to meet any of the requirements for drug substance and drug product CMC information, including alternative components, ingredient sources, manufacturing and packaging procedures, in-process controls, and analytical procedures. If the FDA accepts those alternatives, they may be used to meet CMC requirements. Reference to the current edition of the U.S. Pharmacopoeia and the National Formulary may satisfy relevant requirements in this section.

Also to be included in the NDA is the proposed or actual master production record, including a description of the equipment to be used for the manufacture of a commercial lot of the drug product, or a comparably detailed description of the production process for a representative batch of the drug product.

The FDA's *Draft Guidances*, "Guidance for Industry: Drug Substance Chemistry, Manufacturing, and Controls Information" dated January 2004 and "Guidance for Industry: Drug Product Chemistry, Manufacturing, and Controls Information" dated January 2003, contain detailed information that is recommended to be included in the CMC section of all NDAs. The complete contents of these documents would be beneficial for any NDA applicant to review carefully and thoroughly. Although they are lengthy documents and extremely detailed, they describe the items required for the *drug substance* and the *drug product* related to CMC. The draft guidances include the required granularity for electronic common technical document (eCTD) submission of the CMC section in an NDA. More information on eCTD history and submission requirements is described in the section entitled "Submitting a New Drug Application (NDA)."

MEETINGS WITH THE FDA

Communicating with the FDA during the drug development process can be valuable to individuals or organizations seeking to streamline processes and increase the likelihood of their applications being approved. Although informal communication with the FDA is occasionally appropriate when sponsors have developed a relationship with the FDA review team assigned to their submission, formal communication with

TABLE 3.1
Meeting Formats – Advantages and Disadvantages

Meeting Format	Advantages	Disadvantages
Written Response Only (WRO)	Often completed on a shorter timeline than written responses when a meeting is planned.	Does not allow dialogue with the FDA for sponsors to provide clarification or ask follow-up questions.
Teleconference/ Videoconference	Sponsors do not need to travel to engage in a dialogue with FDA and have an opportunity to ask additional questions and provide further clarification.	Teleconferences do not allow for sponsors and FDA reviewers to meet personally or observe body language as you would in person. Videoconferences may not provide the same opportunity for group interactions and often come with technical issues that interfere with meeting productivity.
Face-to-Face Meeting	Sponsors have an opportunity to personally meet FDA reviewers and engage in dialogue regarding follow-up questions or ongoing concerns from either party.	Sponsors and any representatives that will benefit from being physically present with FDA reviewers must travel to the FDA, which is an additional cost and time investment.

the FDA is generally preferred in order to obtain formal documentation from the FDA regarding specific questions from sponsors. A formal meeting request submitted to the FDA may be for a written response only (WRO), teleconference/video conference, or a face-to-face meeting. In each of these situations, sponsors will submit a list of questions to the FDA regarding their planned submission prior to the meeting with the FDA. FDA will then provide written feedback to the sponsor prior to the meeting, if one is requested. All of these formats have advantages and disadvantages, which have been summarized in Table 3.1. At the time of publication, the FDA had only issued a *Draft* Guidance on Formal Meetings Between the FDA and Sponsors of Applicants of Prescription Drug User Fee Act (PDUFA) Products. The term *PDUFA Products* refers to the Prescription Drug User Fee Act passed in 1992, allowing the FDA to collect fees to fund the new drug approval process. These are any drug or biologic product for which a company or organization submits an NDA or BLA application. The draft guidance on formal meetings between the FDA and sponsors outlines procedures for requesting, preparing, scheduling, conducting, and documenting meetings with the FDA, referred to as good meeting management practices (GMMPs).

Types of Meetings

There are four types of meetings a sponsor can request with the FDA. Depending on the type of meeting, the submission requirements and timelines may vary. Type A meetings are intended to address situations in which product development has stalled or to address a specific safety concern raised by FDA or identified by the sponsor. Sponsors should contact the FDA prior to submitting a request for a Type A meeting to determine the appropriateness of the request. Examples of Type A meetings include clinical hold issues, post-action meetings within three months of an FDA regulatory action other than approval, and dispute resolution described in 21 CFR 10.75, 312.48, and 314.103. Type B meetings include pre-investigational new drug application (pre-IND) meetings, pre-emergency use authorization (EUA) meetings, pre-new drug application (pre-NDA)/pre-biologics license application (pre-BLA) meetings, post-action meetings three or more months after an FDA regulatory action other than approval, and risk evaluation and mitigation strategy meetings prior to the submission of a marketing application. Type B end-of-phase (EOP) meetings are for products that have completed Phase 1, Phase 2, or Phase 3 clinical trials and seek to obtain additional feedback from the FDA regarding their clinical trial plans for the next phase of human studies and any additional preclinical studies planned. Typically, EOP-1 meetings are held only for products that are considering a marketing application under 21 CFR Part 312, subpart E, or 21 CFR 314, subpart H. A Type C meeting is anything that does not fit into the other categories but is related to drug development or FDA review of the product.

Content of Meeting Requests

Enough information must be submitted in a meeting request for the FDA to determine if the requested meeting is appropriate and to identify the FDA staff needed to review the meeting package and discuss the proposed agenda.

Meeting requests *must* include the following information:

1. The proposed meeting format.
2. The date the meeting package will be submitted (see the section entitled “Meeting Package” for more details).
3. A brief statement of the purpose of the meeting. This should include any relevant background information related to how the meeting fits into the general product development plan and a summary of any completed or planned clinical trials or preclinical studies.
4. Specific objectives or outcomes the requester hopes to gain from the meeting.
5. A proposed agenda, including estimated times needed for each item. This should include time for introductions at the beginning and a discussion of next steps at the end. It is important to know how much time you are given, which may vary depending on the type of meeting requested.
6. A list of attendees planned including names and titles, as well as affiliations for any consultants and interpreters, if applicable.
7. A list of FDA attendees or disciple representative(s), if known.

The meeting request *should* include the following information:

1. The application number (if previously assigned).
2. The product name.
3. The chemical name, established name, and/or structure.
4. The proposed regulatory pathway [e.g., 505(b)(1), 505(b)(2)].
5. The proposed indication(s) or context of product development.
6. The meeting type being requested [e.g., Type A, Type B, Type B (EOP), or Type C].
7. Pediatric study plans, if applicable.
8. Human factors engineering plan, if applicable.
9. Combination product information (e.g., constituent parts, including details of the device constituent part, intended packaging, planned human factors studies), if applicable.
10. Suggested dates and times (e.g., morning or afternoon) for the meeting that are consistent with the appropriate scheduling time frame for the meeting type being requested. Dates and times when the requester is not available should also be included.
11. A list of proposed questions, grouped by FDA discipline (e.g., chemistry, manufacturing, and controls (CMC), pharmacology/toxicology, clinical pharmacology and biopharmaceutics, and clinical investigations). For each question, there should be a brief explanation of the context and purpose of the question. This is perhaps one of the most important items included in a meeting request and should be carefully considered before the meeting request is submitted. The questions focus the discussion and should be limited to what the FDA may reasonably review and respond to in the proposed time frame. FDA typically only grants one meeting for each type of Type B meeting, with the exception of products given breakthrough therapy designation. Optimizing timing and content of meetings is critical to successful product development.

Meeting Request Granted or Denied

The FDA will notify the requester in writing if a meeting request is granted or denied. Table 3.2 shows the FDA timeline for responding to meeting requests. If the meeting is granted, the letter may include a date, time, FDA attendees, meeting format, and details of the meeting location. The FDA has to coordinate schedules with all FDA attendees, which does not always allow for one of the dates and times requested in the meeting request to be used. If none of the requested dates and times are available for FDA attendees, the FDA will work with the requester to determine a date and time that works best for the Agency and the requester. The FDA may also grant a meeting in a different format than requested if they deem appropriate, commonly providing WRO when a teleconference or face-to-face meeting was requested. If WRO was requested, but the FDA believes a teleconference or face-to-face meeting is more appropriate, it will notify the requester when granting the meeting. If a meeting is denied, the communication from FDA will include an explanation as to why the meeting was denied. Meeting denials are typically due to significant reasons

TABLE 3.2
Timeline in Calendar Days from Receipt of Meeting Request (when Requester Received Acknowledgement of Receipt)

Meeting Type	Response Time (Grant/Deny)	Meeting Scheduling/WRO Response Time
A	14 days	30 days
B	21 days	60 days
B (EOP)	14 days	70 days
C	21 days	75 days

TABLE 3.3
Timeline in Calendar Days *before* the Scheduled Date of the Meeting/WRO Response

Meeting Type	Meeting Package Submission Time – Days before Scheduled Meeting/ WRO Response
A, C*	At the time of meeting request submission
B	30 days
B (EOP)	50 days
C	47 days

* *For Type C meetings that are requested as early consultations on the use of a new biomarker. All Type A and Type C meetings on early consultation for new biomarker development should include the entire meeting package in the meeting request submission. If the full submission is not received by the FDA at the time of the request, the meeting will likely be denied, and the requester will need to resubmit the request with the complete information.*

such as an insufficient basis for the meeting based on the development stage of the product rather than minor omissions in the meeting request. Table 3.3 shows the FDA timeline before the scheduled date of the meeting/WRO response.

Meeting Package

The required content of the meeting packages will vary depending on the product, indication, stage of development, and current meeting objectives. Meeting packages should be organized in a manner that makes it easy for FDA to navigate through, including numbering pages and bookmarking sections and tables as appropriate. The submission should include a table of contents, indices, appendices, and references. The following information should be included in the meeting package:

The meeting request *should* include the following information:

1. The application number (if previously assigned).

TABLE 3.4
Refuse to File Actions – Examples

Item	Regulation	
1	21 CFR 314.50 21 CFR 601.2(a) 21 CFR 54.4(c) (Financial Disclosures) 21 CFR 314.101(d)(1) through (9)	Materially lacking or inadequately organized applications that would not permit timely, efficient, and complete review by all relevant review division disciplines as outlined in the guidance for review staff and industry *Good Review Management Principles and Practices for PDUFA Products*. This may include data tables or figures that cannot be interpreted or are inadequately labeled, including missing the source of the data; inadequate annotation for preclinical and clinical reports; and issues with hypertext links.
2	21 CFR 314.101(d)(1) through (9)	Parts of applications that contain inadequate information for one or more indications when multiple indications are submitted in the same application. The FDA may accept for filing those parts of an application that represent complete submissions for particular indications but refuse to file those parts that are determined to be incomplete for other indications.
3	Section 505(d) of the FD&C Act	An application that relies on a single adequate and well-controlled trial for a demonstration of effectiveness if prior communication between the FDA and the applicant (i.e., end-of-phase 2 meeting) determined the need for more than one trial to demonstrate effectiveness, and if the submitted justification for reliance on a single trial is inadequate. The FDA may determine that data from a single adequate and well-controlled trial and confirmatory evidence are sufficient to establish effectiveness. This is further discussed in the guidance for industry *Providing Clinical Evidence of Effectiveness for Human Drug and Biological Products*.
4	21 CFR 314.50(d)(5)(vii)	Failure to submit an assessment of studies related to the potential abuse of a drug, necessary to inform drug scheduling under the Controlled Substances Act and the development of drug product labeling. This requirement applies when the drug affects the central nervous system (as determined in animal or human studies), is chemically or pharmacologically similar to other drugs with known abuse potential (such as an opioid, stimulant, depressant, or hallucinogen), or produces psychoactive effects such as euphoria, mood changes, depression, or hallucinations.
5	Section 745A(a) of the FD&C Act (Electronic Submissions)	Required content is not submitted electronically where the FDA has specified the format of such submissions in guidance pursuant to section 745A of the FD&C Act or required content is not submitted in an electronic format that the FDA can review, process, and archive, where such electronic submissions are required by an applicable regulation.
6		NME NDAs or original BLAs reviewed under the Program, if the minor components agreed upon for late submission at the presubmission meeting are not received within 30 calendar days after receipt of the application.

The information in this table is adapted from FDA Draft Guidance "Refuse to File: NDA and BLA Submissions to CDER Guidance for Industry," December 2017.

2. The product name.
3. The chemical name, established name, and/or structure.
4. The proposed regulatory pathway (e.g., 505(b)(1), 505(b)(2)).
5. The proposed indication(s) or context of product development.
6. The dosage form, route of administration, and dosing regimen (frequency and duration).
7. Pediatric study plans, if applicable.
8. Human factors engineering plan, if applicable.
9. Combination product information (e.g., constituent parts, including details of the device constituent part, intended packaging, planned human factors studies), if applicable.
10. A list of attendees planned including names and titles, as well as affiliations for any consultants and interpreters, if applicable.
11. A background section that provides a brief history of product development, communications with the FDA prior to the meeting (i.e., email communications, previous meetings with minutes if available, and previous written responses), and the current status of the product development.
12. A brief statement of the purpose of the meeting and the type of meeting.
13. A proposed agenda, including estimated times needed for each item. This should include time for introductions at the beginning and a discussion of next steps at the end. It is important to know how much time you are given, which may vary depending on the type of meeting requested.
14. A list of final questions grouped by FDA discipline. For each question, there should be a brief explanation of the context and purpose of the question. This is perhaps one of the most important items for any meeting with the FDA. The questions focus the discussion and should be limited to what the FDA may reasonably review and respond to in the proposed time frame. Additional questions asked during the meeting will most likely not be addressed, as the FDA may not have enough background information nor had adequate time to prepare a response.
15. Data to support the questions in Item 14.

Note that some content in the meeting package is included in the meeting request, albeit briefly. Although meeting packages should be concise and information summarized, when possible, the information should be detailed enough for the FDA to assess and respond to the question(s) asked. The goal of the meeting request is to convince the FDA that the meeting is necessary. The meeting package is intended to provide the FDA with the information they will use during the meeting and in their written response. Since the meeting package is used in preparation for the meeting or WRO, it should include any changes to the original meeting request items, including organization attendees or consultants, requested FDA attendees, and the proposed agenda or meeting objectives. Meeting packages should be submitted via the appropriate center's document room or via the electronic submission gateway (ESG), as applicable. Although some guidance documents encourage paper copies to also be provided for all electronic submissions, the FDA is moving away from that practice, and requesters should confirm at the time of submission how many, if any, paper copies the FDA is requesting.

Timeline

It is the responsibility of the sponsor to maintain records of all communication with the FDA, including informal communication before and after meetings. No later than five (5) calendar days before the meeting for Type B (EOP) and Type C, and no later than two (2) calendar days before all other meetings, the FDA will send preliminary responses to the questions in the meeting package. For Type B (EOP) and Type C meetings, sponsors or requesters should notify the FDA acknowledging receipt of the preliminary responses and indicate (1) whether or not the meeting is still needed and (2) a revised meeting agenda indicating any resolved questions and an updated timeline for the remaining questions. Updated agendas should not include any new questions for the FDA or new information not included in the meeting package. Although clarifications may be made and additional discussions may be had at the meeting, the FDA may not be able to comment on new questions or new information.

If the sponsor or requester needs to cancel a meeting for reasons other than the resolution of all questions they had, a new meeting request would need to be submitted to the FDA, and the timeline for the FDA responding and scheduling the meeting would start over. If a meeting needs to be rescheduled, it should be as soon as possible after the original meeting date. When possible, sponsors or requesters should attempt to find a substitute for the meeting if an individual can no longer attend. Any requests for changes to meetings should be communicated to the FDA project manager assigned to the meeting. If the FDA needs to reschedule a meeting, they will communicate that to the individual identified on the meeting request as the appropriate representative.

Meetings with the FDA will begin promptly at the start time and end on or before the end time. An FDA staff member will chair the meeting, manage introductions from FDA, triage responses from FDA, and ensure timeliness of the meeting. A member of the sponsor or requester's organization, or an appropriate consultant, should be prepared to lead introductions of all other attendees, manage the agenda, keep focus on priority items, and keep track of time spent on each question or issue. Someone from the sponsor organization should take detailed minutes of the meeting for internal purposes. Audio or visual recordings of meetings with the FDA are not allowed and are against FDA policy. FDA will issue official meeting minutes within 30 calendar days after the meeting. The minutes will include a summary of discussions, resolutions made, and any action items. FDA may choose to include additional clarifications or information in the meeting minutes for the requester. If the requester does not agree with something in the meeting minutes, or believes there are inaccuracies, the requester should contact the FDA project manager as soon as possible after receiving the meeting minutes. FDA will be the final arbiter on whether or not to make an addendum to the minutes or to reject the requester's concerns and finalize the official minutes as is.

Pre-IND Meetings

Pre-IND Meetings are a Type B meeting and will follow the anticipated timelines described in Tables 8.2 and 8.3. Meetings with the FDA can minimize uncertainty in the development process, provide an opportunity for the FDA to answer questions not specifically addressed in current FDA guidances, and help sponsors avoid clinical hold issues during an IND submission. Many small businesses and researchers new to drug development are often unfamiliar with regulatory processes, do not yet have a vetted way of assessing readiness for different stages of development, and do not have the experience of knowing or working with the FDA. It would be reasonable for companies or individuals with little experience to rely more heavily on the pre-IND meeting process and early communication with the FDA to gain valuable information regarding product development than a pharmaceutical company that has been around as long as the FDA has. Many established companies develop relationships with individuals and review teams within different FDA divisions and have reputations for understanding regulations and following appropriate procedures. These companies may find it beneficial to utilize the pre-IND meeting process to engage the FDA regarding new drug entities, novel indications or drug targets, indications or drug targets without any current guidance from the FDA, or in situations where there may be concern with preclinical data.

Pre-IND meetings can provide useful information to sponsors regarding additional preclinical studies needed before starting human clinical trials, additional manufacturing and control data for the drug product, and/or feedback on study design and development plans. Preclinical studies are expensive and often take additional resources and time. Feedback on CMC data may include additional stability testing, a stability testing plan for during clinical trials, purity/impurity data, or analytical testing procedures and acceptance criteria prior to initiating the planned clinical trial. By obtaining input from the FDA on the type of study and appropriate study design for preclinical studies, as well as what CMC data is required for the drug product, sponsors can avoid clinical hold issues on an IND submission. In fact, inadequate CMC information and insufficient preclinical data are common issues identified in pre-IND meetings. Other issues commonly identified are clinical trial design, including study endpoints, monitoring, and follow-up; non-compliance with GCP; and a lack of data to support the proposed dosage of the drug product. Especially when a drug has never been used in humans before (first-in-human), the dosage proposed by the sponsor should be justified using preclinical data. If preclinical studies were not conducted in a manner that provides sufficient information to support the proposed dosage, sponsors may need to repeat or redesign additional preclinical studies. Sponsors may also get feedback on accelerated approval programs that may be beneficial to product development and strategy. Accelerated approval programs include orphan drug designation, breakthrough therapy designation, fast track designation, and accelerated approval. A company developing a product that may be appropriate for one of these pathways should ensure their product development program is designed in a manner that will provide sufficient supporting information to obtain the desired designation or approval when submitted. The FDA can provide input and guidance on what type of data and endpoints would support such a request. Early communication and transparency with the FDA builds a foundation for open communication and collaboration between sponsors and the Agency.

Pre-NDA/Pre-BLA Meetings

Pre-NDA and Pre-BLA meetings are considered Type B meetings. This is usually one of the last, if not the last, formal meeting between the FDA and a sponsor prior to the submission of a marketing application for a drug (NDA) or biologic (BLA). Pre-NDA meetings help sponsors understand submission readiness and prepare the Agency for the review of the final application. Since the FDA is tasked with reviewing large amounts of data in a relatively short period of time, this type of interaction is helpful to identify any current concerns and unresolved issues and to expose the FDA review team to the information the sponsor intends to submit in the marketing application. This also gives the Agency an opportunity to communicate those concerns with the sponsor so that they are not surprised by any delays in the review process. Common areas of discussion include clinical trial data critical to showing drug efficacy in the proposed indication, appropriateness of the statistical analysis plan, formatting and presentation of important data, and the plan for pediatric safety and efficacy studies.

The pre-NDA or pre-BLA meeting should occur about six months – and no less than 60 days – prior to the planned marketing application submission date. Depending on the sponsor and their preferred approach toward compiling regulatory submissions, this type of meeting may be held before or after the sponsor starts to assemble the marketing application. The Agency does not require the marketing application to be submission-ready to hold a meeting; however, any feedback the sponsor hopes to receive regarding the format of data in the submission, as well as the format and organization of the submission itself, may be limited to what the sponsor provides to the FDA or is prepared to discuss with the FDA. Pre-NDA/Pre-BLA meetings typically include a discussion of the following:

1. Proposed labeling and data to support what is included in the labeling.
2. Integrated summary of safety (ISS) and integrated summary of efficacy (ISE) data.
3. Outstanding issues from previous meetings with the FDA that have not already been addressed or resolved.
4. Plans for manufacturing, testing, and packaging of the drug product intended for marketing, compared to what was done in Phase 3 studies. Facilities for the marketed product should be inspection-ready at the time of NDA/BLA submission.
5. Early discussions of risk management and post-marketing studies.
6. Discussions on requests for Priority Review, Accelerated Approval, or Rolling Review.
7. Any additional concerns that may cause delays in the NDA/BLA review.

INVESTIGATIONAL NEW DRUG APPLICATIONS (INDS)

Investigational New Drug Applications (INDs) are submitted to the FDA when a sponsor of a clinical trial wants to administer an investigational drug or biological product to human subjects. Federal law prohibits the transportation and distribution of a drug across state lines without an approved marketing application, such as an NDA or BLA. The IND process is the manner in which sponsors obtain an exemption for this requirement in order to distribute a drug for clinical trials. This stage in drug development usually comes after a sponsor has conducted preclinical safety and toxicology studies *in vitro* and in animal models. When necessary, sponsors should request Pre-IND meetings with the FDA to review the current and planned preclinical studies along with the proposed human study protocol. This will optimize the likelihood of their IND application being approved. Once an IND application is approved, allowing for the investigational use of the drug in human subjects as described in the protocol, the drug must comply with several additional regulatory requirements.

There are two categories of IND applications: commercial and research (noncommercial). The category must be selected by the sponsor of the study when the IND application is submitted, indicating the appropriate category on Form FDA 1571. Typically, research INDs are those that are initiated by a physician, researcher, or research organization with no intent to commercialize the investigational drug or use the data to make any changes to the drug labeling. An individual that submits a research IND is also often directly involved with supervising the dispensing and administration of the investigational drug. Commercial INDs are those that are typically conducted by a corporate entity with the intention of eventually commercializing and profiting off the product. Occasionally, academic institutions, individual researchers, or physicians as sponsors will intend to commercialize the investigational drug. In these situations, the IND would be considered a commercial IND rather than a research IND despite the sponsor not being a corporate entity.

There are also three different types of INDs: Investigator IND, Emergency Use IND, and Treatment IND. Investigator INDs are submitted by an individual, usually a physician or established researcher, who initiates and conducts the investigation. These investigations may involve an unapproved drug or an approved drug in an unapproved use, indication, or a new patient population. Emergency Use INDs are an authorization from the FDA to use an experimental drug in an emergency situation when the submission of an IND in accordance with 21 CFR Part 312 is not feasible given the time constraints of the emergency. Treatment INDs are used to treat patients prior to the completion of clinical trials and FDA review, when an experimental drug is being studied for a serious or immediately life-threatening condition(s) and is anticipated to obtain approval.

There are three primary sections of an IND application. First, the preclinical safety, pharmacology, and toxicology studies provide information to the FDA regarding the laboratory and animal data that support the use of the drug product in humans as described in the protocol. This section would also include any previous experience with the drug in humans and may include data from published literature. Second, the chemistry, manufacturing, and controls (CMC) information includes details regarding the manufacturer, composition, manufacturing process, stability, and controls used in the manufacturing of both the drug substance and drug product. This ensures that the drug being administered to human subjects is reliable, stable, and consistently made from batch to batch. The drug substance is also referred to as the active pharmaceutical ingredient (API) and is "intended to have a pharmacological effect or other effect in the diagnosis, cure, mitigation, treatment or prevention of disease, or to affect the structure or any function of the body of man or other animal" (OFR 2019b). This includes compounds that undergo a chemical change in the manufacturing process and are present in the final drug product as the active ingredient. The drug product is the final dosage form that delivers the active ingredient to the study subjects and eventually to patients (i.e., tablet, capsule, patch, suspension, cream, or ophthalmic drops). The drug product may undergo several changes throughout development, some of which may take place during clinical trials. Changes made to the manufacturing process or formulation that impact the drug product should be clearly documented with a justification for the change and submitted to the FDA. The term drug product also refers to a dosage form with no active ingredient but containing excipients and intended for use as a placebo in clinical trials. Investigator-sponsored research INDs often address the nonclinical and CMC information requirements by providing a Letter of Authorization (LOA) from the commercial entity or manufacturer. This allows the FDA to reference that information without the commercial company or manufacturer having to disclose it directly to the researcher. Lastly, the clinical protocol and investigator information, including medical licenses and *curricula vitae* (CVs), must be included in the IND application. The protocol should be detailed sufficiently for the FDA to assess and determine if the procedures will expose study subjects to unreasonable risks. The investigator information submitted should be sufficient for the FDA to determine if they are qualified by training or experience to fulfill the responsibilities of a clinical investigator. Sponsors are also required to commit to obtaining informed consent and approval from an institutional review board (IRB) prior to study initiation. The FDA may request documentation of such, and the sponsor should ensure all requests from the FDA are fulfilled.

Aside from the three primary sections of an IND application that support the use of the product in humans, the submissions also contain technical requirements such as a Cover Letter, Forms FDA 1571 Investigational New Drug Application, Form FDA 1572 Statement of Investigator, and Form FDA 3674 Certification of Compliance with Requirements of the ClinicalTrials.gov Data Bank. Beginning in 2018, all commercial IND applications must be submitted using the electronic common technical document (eCTD) format. The eCTD format is encouraged for research IND submissions, but it is not required at this time. For research IND submissions opting out of using the electronic submission system and format, they may submit the application via paper submission and should follow the requirements for such. INDs submitted with paper applications should not intend to commercialize the investigational product and typically report their research findings in peer-reviewed journals. Once a complete IND application is submitted to the FDA, the sponsor must wait 30 calendar days before initiating any clinical trials. During this review period, the FDA reviews the IND for any potential safety concerns that would put study subjects at unnecessary risk. If the sponsor does not hear from the FDA within 30 days, it is reasonable to proceed with the study as planned. The FDA typically sends a Study May Proceed letter to sponsors if or when their IND is approved.

CLINICAL STUDY PHASES UNDER AND IND

It is imperative that clinical trials be designed, conducted, and analyzed using sound scientific principles. The studies should be designed in such a way as to answer appropriate questions regarding a new drug's safety and efficacy. Designing clinical trials to be conducted in phases allows for the results of the early studies to influence the development of later studies. The data generated from each clinical study will impact the ongoing product development plans. The FDA's primary objectives in reviewing an IND are, in all phases of the investigation, to assure the safety and rights of human subjects, and, in Phases 2 and 3, to assure that the quality of the scientific evaluation of drugs is adequate to permit an evaluation of the drug's effectiveness and safety.

The FDA encourages IND sponsors to engage with the agency to resolve questions and issues raised during the course of a clinical investigation. This is done through a formal meeting process, which is described in detail in the section entitled "Meetings with the FDA." Meetings regarding INDs are free, and there should be open communication about any scientific or medical question(s) that may arise during clinical investigations. Meetings between the FDA and a sponsor can be especially helpful in minimizing wasteful expenditures of time and money and thus in speeding the drug development and evaluation process. The clinical investigation of a previously untested drug is generally divided into three phases and one phase post-approval. Although the phases are generally conducted sequentially, they may overlap.

Phase 1

Phase 1 studies are primarily conducted in healthy human subjects with the goal of establishing the safety profile of the drug at a pre-specified dose range, which will guide dosing regimens for future studies and inform the sponsor of potential safety concerns or side effects. There are exceptions to this when, for example, the drug is toxic, such as oncologic drug products. Phase 1 studies also include studies of drug metabolism, structure-activity relationships, and mechanisms of action in humans, as well as studies in which investigational drugs are used as research tools to explore biological phenomena or disease processes.

> Phase 1 includes the initial introduction of an investigational new drug into humans. Phase 1 studies are typically closely monitored and may be conducted in patients or normal volunteer subjects. These studies are designed to determine the metabolism and pharmacologic actions of the drug in humans, the side effects associated with

> increasing doses, and, if possible, to gain early evidence on effectiveness. During Phase 1, sufficient information about the drug's pharmacokinetics and pharmacological effects should be obtained to permit the design of well-controlled, scientifically valid, Phase 2 studies. The total number of subjects and patients included in Phase 1 studies varies with the drug but is generally in the range of 20 to 80. (OFR 2019d)

Phase 1 Professional Research Subjects

Because Phase 1 clinical trials are often conducted with healthy subjects, one must keep in mind that some of these "healthy subjects" may be "professional research subjects." These are individuals who make a living by participating in these studies. A full-time "human guinea pig," as they are often referred to, can make $15,000 to $40,000 a year (Romm 2015).

Aside from the fact that the "human guinea pigs" may be damaging their own bodies, they may be putting entire studies in jeopardy. If professional guinea pigs get caught in a lie, they can either get removed from a study, or it can cause a company to have to throw out data or scrap an entire study because of unreliable data. This can potentially cost an IND sponsor millions of dollars. On the ethical side, the data from legitimate subjects who participated may be scrapped as well. These healthy volunteers may have been exposed to an investigational drug for no benefit to anyone.

Phase 2

Phase 2 studies are efficacy studies generally conducted for dose regimens that are anticipated to be safe based on Phase 1 studies. Sometimes Phase 2 studies are conducted as Phase 2A and Phase 2B studies. Phase 2A studies are usually focused on determining dosing requirements for the proposed efficacy and mechanism of action of the drug. Phase 2B studies are conducted to evaluate the efficacy of the drug at different doses, and Phase 3 contains pivotal safety and efficacy studies.

> Phase 2 studies evaluate the effectiveness of the drug for a particular indication or indications in patients with the disease or condition under study and to determine the common short-term side effects and risks associated with the drug. Phase 2 studies are typically well controlled, closely monitored, and conducted in a relatively small number of patients, usually involving no more than several hundred subjects. (OFR 2019d)

Phase 3

Phase 3 studies are large studies to confirm the drug's effectiveness, monitor side effects, compare with other treatments and/or placebo, and collect other information that will support the safe and effective use of the drug for the indication for which it is studied.

> Phase 3 studies are expanded controlled and uncontrolled trials. They are performed after preliminary evidence suggesting effectiveness of the drug has been obtained and are intended to gather the additional information about effectiveness and safety that

> is needed to evaluate the overall benefit-risk relationship of the drug and to provide an adequate basis for physician labeling. Phase 3 studies usually include from several hundred to several thousand subjects. (OFR 2019d)

Phase 4

Phase 4 clinical studies are conducted post-marketing, that is, after the NDA has been approved. They are often referred to as post-marketing surveillance studies. The commitment to Phase 4 studies is often a condition in the approval letter of an NDA. Often, the FDA will require the sponsor to conduct trials after approval to obtain additional information on the safety of the drug, pharmacovigilance, and how the drug is used in the practice of medicine as opposed to in a controlled clinical trial. If harmful effects are discovered, this can result in the drug being pulled off the market or restricted to certain uses and patient populations. One example is the pain reliever rofecoxib, which showed negative cardiac side effects during Phase 4 trials and was therefore withdrawn from the market (Mandal 2019).

Because these drugs are being lawfully marketed, placebo-controlled studies are generally ruled out for ethical reasons. The FDA takes these studies seriously and may issue warning letters to sponsors who neglect to conduct or finish the studies. A sponsor that fails to respond to a Phase 4 commitment may receive an administrative action letter, also referred to as a "Dunner" letter, from the FDA that states: "Our letter of DATE [, approving this application,] notes the above Phase 4 commitment[s]. To date, we have not received any/a complete response to this[these] commitment[s]. We request that you submit your plan for fulfilling your Phase 4 commitments, or your rationale for why you should be released from this [these] commitment[s]" (USFDA 1996).

The Food and Drug Administration Amendments Act (FDAAA) of 2007 requires that an applicant report periodically on the status of any study or clinical trial required for Phase 4. Failure to submit an annual report for studies or clinical trials required under section 505(o) of the FDCA on the date required will be considered a violation of FDCA section 505(o)(3)(E)(ii) and could result in enforcement action (USFDA 2018).

SUBMITTING A NEW DRUG APPLICATION (NDA)

The importance of a complete and properly formatted NDA cannot be overstated. Hundreds of millions of dollars, many years, and an extreme amount of resources have been poured into the development of a new drug. Pharmaceutical company executives are expected to bring new drugs to market. Stockholders expect an increase in the equity of their shares. Patients depend on new drugs to treat their ailments and help improve their quality of life. The technical requirements for submitting a new drug application must be followed and submitted correctly for the FDA to accept the application and complete their review.

The Electronic Common Technical Document (eCTD)

For decades, NDAs were submitted to the FDA in triplicate on paper. Each copy of the NDA consisted of scores of volumes. Each volume was about 3 inches thick. These paper copies were stored in the FDA's Document Control Center (DCC). In order for NDA reviewers to review their assigned sections, the appropriate volumes were sent to the reviewer. Other reviewers who needed those same volumes had to wait for them to be returned to the DCC. This was a tremendous waste of time, paper, and storage space.

Even though the regulation 21 CFR 314.50 specifies the content and format of an NDA, there were inconsistencies from one application to another and from sponsor to sponsor. In 2015, the FDA provided a timeline for the initial implementation of the electronic submission requirement for NDAs (May 5, 2017), for commercial INDs (May 5, 2018), and for Drug Master Files (DMFs) (May 5, 2019). In section 745A(a) of the FD&C Act, Congress granted authorization to the FDA to implement the electronic submission requirements.

The FDA created the Common Technical Document (CTD) format. All applications were required to be submitted in electronic CTD format through the FDA's Electronic Submission Gateway (ESG). The format of the eCTD consists of five modules (1. Administrative Information; 2. Summaries; 3. Quality; 4. Nonclinical Study Reports; 5. Clinical Study Reports). The content of each module is broken down into many separate files, referred to as document granularity. This granularity must be consistent with the Granularity Document found in the ICH guidance for industry *M4 Organization of the electronic common technical document for the registration of pharmaceuticals for human use* (USFDA, CDER, CBER 2020).

The creation of the eCTD made the format of the contents the same for every NDA. The NDA was in electronic format. Each reviewer had full-time access to the entire NDA. This made the NDA review process more efficient, which also shortened the review time.

Contents of an NDA

A New Drug Application (NDA) is perhaps the most important compilation of data for a drug. Once a drug is approved, the contents of the NDA are made available to the public with the exception of proprietary and confidential information. The data accumulated from drug discovery to *in vitro* studies to preclinical safety and toxicology studies to clinical trials; manufacturing information and scalability; quality control and release criteria; and package labeling with directions for use are all summarized in the NDA. By submitting an NDA, the drug developer is claiming that the drug is safe and effective for the intended use and will be able to be used appropriately by prescribers and/or patients when used according to the package labeling. They should provide enough information for the FDA to determine whether the benefits of the drug outweigh the risks, including any data relevant to making that assessment, such as drug abuse information and proposed safety monitoring plans, if applicable.

The first step of the FDA review of an NDA submission is similar to that of meeting requests and IND applications: an objective assessment by the review team of the NDA's completeness. If the NDA is complete, the FDA has 180 days to decide whether or not to approve the drug. The "filing date" of the NDA will be 60 days after the date the FDA received the NDA. The filing date begins the 180-day period described in section 505(c) of the Federal Food, Drug, and Cosmetic Act.

The following information and documents must be included in an NDA application:

1. Form FDA-356h: "Application to Market a New Drug, Biologic, or An Antibiotic Drug for Human Use." This form is typical to other applications to the FDA and include basic information, applicant name and address, product details, type of submission, manufacturer information, contents of the submission, and official signature.
2. Form FDA-3397: "User Fee Cover Sheet." FDA is authorized to charge for PDUFA submissions under sections 735 and 736 of the Federal Food, Drug, and Cosmetic Act (21 U.S.C. 379g and 379h). This may be submitted online separately, and documentation may be provided in the NDA. User fees vary depending on the type of application and type of data being submitted to the FDA, but as of FY 2022, the fee was over three million U.S. dollars (USD) for applications requiring clinical data, due at the time of filing the application.
3. Form FDA-3331 "New Drug Application Field Report."
4. Environmental assessment (EA) per 21 CFR 25.40 or information to establish a categorical exclusion under 21 CFR 25.30 or 21 CFR 25.31 (§ 314.101(d)(4)).
5. Accurate and complete English translation of each part of the NDA that is not in English (§ 314.101(d)(5)).
6. Index and table of contents (§ 314.50(b)).
7. Summary of the application (§ 314.50(c)).
8. Chemistry, manufacturing, and controls (§ 314.50(d)(1)).
    - Applicants should provide adequate information that assures identity, strength, purity, and quality of the drug substance or drug product.
    - Complete information (i.e., name, address, and contact information) of all facilities involved in the manufacturing process (e.g., drug substance and drug product, control and testing labs, primary packaging and labeling).
    - All manufacturing sites intended for production of the to-be-marketed drug product should be registered with the FDA at the time the application is submitted.
    - Facilities should be inspection-ready at the time the application is submitted.
    - The application should specify what responsibilities each facility has in supporting the manufacturing, packaging, and labeling activities described in the application.

- All impurities should be characterized to the extent possible and appropriate toxicology studies conducted to address them.
- Stability data should be submitted to support a commercially viable expiration dating period.

9. Nonclinical pharmacology and toxicology (§ 314.50(d)(2)).Each nonclinical laboratory study should include a statement that the study was conducted in compliance with the requirements set forth in 21 CFR part 58 or, for each study not conducted in compliance with part 58, a brief statement of the reason for the noncompliance (§ 314.101(d)(6)).
    - Carcinogenicity studies must be included for a drug product intended to be administered chronically or an adequate explanation of why the studies are not necessary.
    - Reproductive toxicology studies must be included for a drug product intended for use in people of reproductive age or an adequate explanation of why the studies are not necessary.
    - Any limits on impurities, as defined by the International Conference on Harmonisation, should not be exceeded without accompanying animal studies to evaluate the safety of these impurities.
10. Human pharmacokinetics and bioavailability (§ 314.50(d)(3)).
    - For a 505(b)(2) application, comparative bioavailability data between the proposed drug product and the reference drug product should be submitted to demonstrate that such reliance is scientifically justified. The referenced drug product must be an approved drug product at the time the application is filed.
    - Bioequivalence data comparing the to-be-marketed drug product with the drug product used in the pivotal clinical trials should be submitted (e.g., bridging trials that support the marketed formulation).
    - Bioanalytical method validation and trial-specific bioanalytical method performance information for the bioanalytical assays used to determine drug concentrations in biological matrices.
    - Bioavailability data or a request for biowaiver.
    - Drug disposition information.
    - Drug-drug interaction information.
11. Microbiology, if the drug is anti-infective (§ 314.50(d)(4)).
12. Clinical data (§ 314.50(d)(5)).
    - Each clinical study should contain a statement that the study was conducted in compliance with the IRB regulations in 21 CFR part 56, or was not subject to those regulations, and that it was conducted in compliance with the informed consent regulations in part 50, or if the study was subject to but was not conducted in compliance with those regulations, a brief statement of the reason for the noncompliance (§ 314.101(d)(7)).
    - Clinical trial protocols should be included with any amendments to the clinical trial design or statistical analysis plan clearly documented.

Amendment summaries should include dates in reference to data locks and data analyses.
- Randomization information such as treatment allocation by site, day, and time; randomization scheme; and randomization ratio.
- For a 505(b)(2) application, literature references or listed drug citation to support the safety/efficacy of the drug product should be included.
- For a combination drug product, studies/clinical trials should assess the contribution of each component, or provide an adequate explanation and supporting data of why the requirement should be waived.
- Required case report forms (CRFs) and tabulations (§ 314.50(f)).
- Evidence of safety and effectiveness compatible with statute and regulations.
    i. Clinical trial design and study endpoints should be appropriate or clinically relevant for the claimed indication(s).
    ii. If primary study endpoints did not meet significance or endpoints are used that may not be known to predict clinical benefit (under 21 CFR part 314, subpart H, and under part 601, subpart E), the applicant should include an explanation or justification of why it should still be considered and may be *reasonably likely to predict* a clinical benefit.
    iii. Patient reported outcomes (PROs) should be validated for the context of the applicable clinical trials or justification of why it should be considered anyways.
    iv. Sufficient data should be collected and evaluated in the population intended to use the drug product, including pertinent subsets, such as sex, age, and racial subsets.
    v. Safety data should be obtained for a sufficient duration of time to evaluate safety in long-term exposure for chronically administered drugs.
    vi. Analysis of data supporting the proposed dose and dose interval.

13. Integrated summary of effectiveness (§ 314.50(d)(5)(v)).
14. Integrated summary of safety (§ 314.50(d)(5)(vi)).
15. Statistical evaluation (§ 314.50(d)(6)).
    - Include critical statistical analyses with adequate justification and explanation.
    - Demographic subset analyses specified in the regulations (§ 314.50(d)(5)(V) and (VI)) should be included.
    - Statistical analysis methods should be conducted and documented in a manner that maintains trial integrity, such as finalizing the statistical plan before finalizing the protocol.
    - Any changes to the statistical analysis plan should be documented in the respective protocols.
16. Pediatric use (§ 314.50(d)(7)).
    - Address requirements under PREA, including a pediatric assessment or an initial pediatric study plan agreed upon.

17. Annotated package insert and the marketing history of the drug product outside the United States (§ 314.50(c)(2)).
18. Labeling content and format per 21 CFR 201.56 and 201.57 (*Requirements on Content and Format of Labeling for Human Prescription Drug and Biological Products*) or in the physician labeling rule format (see the guidance for industry *Labeling for Human Prescription Drug and Biological Products – Implementing the PLR Content and Format Requirements*).
19. Limited distribution strategy for a drug product shown to be effective but that can only be safely used if distribution or use is restricted (21 CFR 314.520).
20. Abuse liability section if the application is one for which this section is required, including a proposal for scheduling under the Controlled Substances Act (§ 314.50(d)(5)(vii)).
21. Risk Evaluation and Mitigation Summary, if applicable.
22. Integrated summary of the benefits and risks of the drug product (§ 314.50(d)(5)(viii)).
23. Patent certification or statement as described under § 314.54(a)(1)(vi) for a 505(b)(2) application relying on one or more listed drugs.

If the application is *incomplete*, the FDA team can *refuse to file* (RTF) the NDA. 21 CFR 314.101(d)(3) states that FDA may refuse to file an NDA if it "is incomplete because it does not on its face contain information required under section 505(b) … of the Federal Food, Drug, and Cosmetic Act and § 314.50." Deficiencies leading to a RTF action are those that render the application unreviewable and are inconsistent with statutory or regulatory requirements. Not all deficiencies will lead to a RTF action, especially minor issues that are easily correctable within the 60-day filing period. More complex or significant deficiencies that are unlikely to be corrected in the 60-day filing period are what lead to a RTF action. Examples of those deficiencies are listed in Table 3.4. There are some exceptions to incomplete applications. As previously discussed, sponsors may hold pre-NDA meetings with the FDA and discuss the contents of the planned NDA submission. In that meeting, FDA and the applicant may agree to the delayed submission of certain minor components of the application; however, all components must be submitted within 30 days of the NDA filing. FDA permitting exceptions to the rules in situations such as this one makes the assessment of an application's completeness more complicated and challenging for FDA review teams. Applicants should ensure that if they are given such leniency by the Agency, they abide by the timeline agreed to. Any deficiencies noted in the application may be communicated to the applicant as soon as they are identified by FDA, allowing the applicant as much time as possible to address deficiencies before FDA's deadline for determinations. FDA commits to reviewing applications on a strict timeline under the Prescription Drug User Fee Act (PDUFA), and any delays in submission of information by the applicant that risk FDA's predetermined timeline will likely not be accepted.

If the FDA determines an RTF action is appropriate, they must notify the applicant in writing within the 60-day filing period. The applicant may then, within 30 days of the RTF notification, request an informal conference with the FDA to discuss whether the FDA should file the application. The applicant may determine that they will amend the NDA and resubmit it at a later date, which will undergo an independent filing and review process. The applicant may also request that the FDA review division file the application despite the Agency's concerns, in which case the review division will file the application "over protest" pursuant to 21 CFR 314.101(a)(3). Once an application is filed, the FDA may then encounter filing issues. If there are any filing issues, the applicant will be notified in writing by the FDA within 14 days after the 60-day filing date. The FDA will convey all filing issues and the Agency's expectations for applicant response(s) in a single communication.

Once the application has been filed, the FDA will begin the 180-day review period. Each person on the review team assigned will review their section of the application. Each technical discipline involved in reviewing the application also has a supervisory review after the initial review. FDA inspectors will travel to clinical study sites to conduct inspections. Applicants should ensure clinical trial sites and manufacturing facilities are adequately prepared for inspection at the time of NDA submission. Inspections are conducted to identify evidence of fraud or non-compliance with good clinical practice. The FDA may organize an "Advisory Committee" of independent, external experts and patient representatives to provide additional input on drug approval if the FDA identifies the need for this additional consideration. The project manager assigned to the NDA review will collate all individual reviews and other documents, such as the inspection report(s), into an "action package." This will include deficiencies identified that require the developer to address concerns with the existing data or processes. For example, deficiencies cited during FDA inspections of clinical trial sites or manufacturing facilities may require action from the applicant prior to FDA approval. Occasionally, the FDA requires additional studies to be performed. If that occurs, the applicant can decide whether or not to continue further development. If a developer disagrees with an FDA decision, there are mechanisms for formal appeal. When the FDA determines that the drug has been shown to be safe and effective for the intended use, there is still often work to be done between the applicant and the FDA prior to the drug being approved for marketing. The FDA will review the proposed labeling for accuracy and user comprehension and will work with the applicant to refine the package labeling and prescribing information if necessary.

PEDIATRIC REQUIREMENTS

One of the most important concepts to learn in drug development and in medicine is that *children are not small adults*. Although there are many challenges in conducting studies in children, it is an important requirement for companies developing drugs that may be used in pediatric populations because you cannot always assume that pediatric medicine is simply using a lower dose than in adults. Children have different body compositions, different metabolic enzymes which impact pharmacokinetic predictions and metabolite concentrations, as well as different pharmacodynamic properties depending on the drug and the indication for use. To the FDA, pediatric patients are usually considered to be children from birth to 16 years of age (OFR 2019c). To an IRB, children are defined as persons who have not attained the legal age for consent to treatments or procedures involved in clinical investigations, under the applicable law of the jurisdiction in which the clinical investigation will be conducted (OFR 2019a).

In the past, pediatricians had few choices regarding drug treatment for children: not treat children with potentially beneficial medications because they were not approved for use in children, or treat children with medications based on adult studies with limited or anecdotal pediatric experience (off-label use). There are several references that address pediatric dosage calculations. To this day, for example, the Lexi-Comp Pediatric & Neonatal Dosage Handbook continues to be a trusted resource for all medical professionals managing pediatric patients.

Several formulas have been developed to assist in determining the proper dose of adult drugs for children (Bradley WT et al., 1963).

- Young's Rule: (Age / Age + 12) x Adult dose = Dose for child
- Cowling's Rule: (Age + 1 x Adult dose) / 24 = Dose for child
- Fried's Rule for Infants: (Age in months / 150) x Adult dose = Dose for infant
- Clark's Rule for Infants: (Weight in pounds / 150) x Adult dose = Dose for infant

Crawford JD et al. published a table for individuals with average body dimensions that approximates the relation of surface area to weight. The rationale for using this method, as stated by the authors, is that "It was believed that doses for children should be based upon body surface area, since the correct dosage of drugs seems nearly proportional to the surface area" (Crawford JD et al. 1950).

In December 2000, the FDA issued a final guidance document for the industry "E11 Clinical Investigation of Medicinal Products in the Pediatric Population." This guidance was prepared under the auspices of the International Conference on Harmonisation of the Technical Requirements for Registration of Pharmaceuticals for Human Use (ICH). The guidance provides an outline of critical issues in pediatric drug development and approaches to the safe, efficient, and ethical study of medicinal products in the pediatric population.

In 21 CFR 314.55 Pediatric use information, it states that each application for a new active ingredient, new indication, new dosage form, new dosing regimen, or new route of administration shall contain data that is adequate to assess the safety and effectiveness of the drug product for the claimed indications in all relevant pediatric subpopulations and to support dosing and administration for each pediatric subpopulation for which the drug is safe and effective. The FDA may, on its own initiative or at the request of the NDA sponsor, defer submission of some or all assessments of safety and effectiveness until after approval of the drug product for use in adults. If the FDA determines that there is adequate justification for temporarily delaying the submission of assessments of pediatric safety and effectiveness, the drug product may be approved for use in adults, subject to the requirement that the applicant submit the required assessments within a specified time. The requirement for satisfying this waiver would be specified in the NDA approval letter. This is an example of a Phase 4 clinical trial.

Today, there are two pediatric development laws:

1. The Pediatric Research Equity Act (PREA) requires companies to assess the safety and effectiveness of new drugs/biologics in pediatric patients (Pediatric Assessment).
2. The Best Pharmaceuticals for Children Act (BPCA) provides a financial incentive to companies to voluntarily conduct pediatric studies of drugs that still have marketing exclusivity.

FDASIA Permanently Reauthorized PREA and BPCA

The FDA Safety and Innovation Act (FDASIA) permanently reauthorized PREA and BPCA. The FDA is serious about the PREA requirements. An NDA sponsor that is not in compliance with PREA requirements may receive a "Notice of non-compliance with PREA" letter. This letter states:

> The Agency has determined that you have failed to meet the post marketing requirements (PMRs) of the Pediatric Research Equity Act (PREA) for these applications because you have not yet submitted your pediatric assessment for PMRs 26-10 and 1784-5, which were deferred until November 30, 2020. Therefore, we are hereby notifying you that due to your failure to submit either a pediatric assessment or a request for a deferral extension, you are not in compliance with federal law. Under the provisions of section 505B(d)(1) of the Federal Food, Drug, and Cosmetic Act (FD&C Act) [21 U.S.C. 355c(d)(1)], you must respond in writing within 45 calendar days of the date of this letter. Your response should include the reason(s) for the delayed pediatric assessment and a date by which you expect to submit the assessment. You may also include a request for a deferral extension, if applicable."

REFERENCES

Abraxis BioScience, LLC. August 2020. "ABRAXANE (paclitaxel) Label." https://www.accessdata.fda.gov/drugsatfda_docs/label/2020/021660s047lbl.pdf.

Bradley WT, et al. 1963. *Pharmaceutical Calculations*. 4th ed. Lea & Febiger, pp. 86–87.

Bristol-Myers Squibb Company (BMSC). April 2011. "Taxol (paclitaxel) Injection Label." https://www.accessdata.fda.gov/drugsatfda_docs/label/2011/020262s049lbl.pdf.

Coronavirus Aid, Relief, and Economic Security Act (CARES Act), Pub. L. No. 116–136, 134 Stat. 281 (2020).

Crawford JD, et al. 1950. "Simplification of Drug Dosage Calculation by Application of the Surface Area Principle." *Pediatrics*, 5, pp. 783–790.

Mandal, Ananya. February 26, 2019. "What is a Phase 4 Clinical Trial?" *News Medical*. https://www.news-medical.net/health/What-is-a-Phase-4-Clinical-Trial.aspx.

Office of the Federal Register (OFR), National Archives and Records Administration. April 1, 2019a. "Code of Federal Regulations." Title 21 Food and Drugs, Volume 1, Part 50.3 - Protection of human subjects: Definitions. https://www.govinfo.gov/content/pkg/CFR-2020-title21-vol1/pdf/CFR-2020-title21-vol1.pdf.

Office of the Federal Register (OFR), National Archives and Records Administration. April 1, 2019b. "Code of Federal Regulations." Title 21 Food and Drugs, Volume 4, Part 210.3 - Current good manufacturing practice in manufacturing, processing, packing, or holding of drugs; general. https://www.govinfo.gov/content/pkg/CFR-2020-title21-vol4/pdf/CFR-2020-title21-vol4.pdf.

Office of the Federal Register (OFR), National Archives and Records Administration. April 1, 2019c. "Code of Federal Regulations." Title 21 Food and Drugs, Volume 4, Part 210.57 - Labeling: Specific requirements on content and format of labeling for human prescription drug and biological products described in § 201.56(b)(1). https://www.govinfo.gov/content/pkg/CFR-2020-title21-vol4/pdf/CFR-2020-title21-vol4.pdf.

Office of the Federal Register (OFR), National Archives and Records Administration. April 1, 2019d. "Code of Federal Regulations." Title 21 Food and Drugs, Volume 5, Part 312.21 - Investigational new drug application: Subpart B—Investigational New Drug Application (IND) - Phases of an investigation. https://www.govinfo.gov/content/pkg/CFR-2020-title21-vol5/pdf/CFR-2020-title21-vol5.pdf.

Office of the Federal Register (OFR), National Archives and Records Administration. April 1, 2019e. "Code of Federal Regulations." Title 21 Food and Drugs, Volume 5, Part 355 - Anticaries drug products for over-the-counter human use. https://www.govinfo.gov/content/pkg/CFR-2020-title21-vol5/pdf/CFR-2020-title21-vol5.pdf.

Romm, Cari. September 23, 2015. "The Life of a Professional Guinea Pig: What It's Like to Earn a Living as a Research Subject in Clinical Trials." *The Atlantic*. https://www.theatlantic.com/science/archive/2015/09/life-of-a-professional-guinea-pig/406018/.

The International Council for Harmonisation of Technical Requirements for Pharmaceuticals for Human Use (ICH). July 12, 2021. "ICH Official web site: ICH Quality Guidelines." https://ich.org/page/quality-guidelines.

United States Food and Drug Administration (USFDA). February 12, 2018. "Postmarketing Requirements and Commitments: Frequently Asked Questions (FAQ)." Question 2. https://www.fda.gov/drugs/postmarket-requirements-and-commitments/postmarketing-requirements-and-commitments-frequently-asked-questions-faq#q2.

United States Food and Drug Administration (USFDA). December 31, 2020. "CDER Drug and Biologic Calendar Year Approvals." https://www.fda.gov/media/147397/download.

United States Food and Drug Administration (USFDA). July 13, 2021a. "Summary of NDA Approvals & Receipts, 1938 to the present." https://www.fda.gov/about-fda/histories-product-regulation/summary-nda-approvals-receipts-1938-present.

United States Food and Drug Administration (USFDA). July 14, 2021b. "Drugs@FDA: FDA-Approved Drugs." https://www.accessdata.fda.gov/scripts/cder/daf/.

United States Food and Drug Administration (USFDA). February 2021c. "New Drugs at FDA: CDER's New Molecular Entities and New Therapeutic Biological Products." https://www.fda.gov/drugs/development-approval-process-drugs/new-drugs-fda-cders-new-molecular-entities-and-new-therapeutic-biological-products.

United States Food and Drug Administration (USFDA). February 2021d. "Small Business and Industry Assistance: Frequently Asked Questions on the Pre-Investigational New Drug (IND) Meeting." https://www.fda.gov/drugs/cder-small-business-industry-assistance-sbia/small-business-and-industry-assistance-frequently-asked-questions-pre-investigational-new-drug-ind.

United States Food and Drug Administration (USFDA). June 1, 2021e. "Facts About the Current Good Manufacturing Practices (CGMPs)." https://www.fda.gov/drugs/pharmaceutical-quality-resources/facts-about-current-good-manufacturing-practices-cgmps.

United States Food and Drug Administration (USFDA), Center for Drug Evaluation and Research (CDER). October 1, 1996. "Manual of Policies and Procedures: Review Management: Procedures for Tracking and Reviewing Phase 4 Commitments." https://www.fda.gov/files/about%20fda/published/Procedures-for-Tracking-and-Reviewing-Phase-4-Commitments.pdf.

United States Food and Drug Administration (USFDA), Center for Drug Evaluation and Research (CDER), Center for Biologics Evaluation and Research (CBER). July 2011. "Guidance for Industry S7A Safety Pharmacology Studies for Human Pharmaceuticals." https://www.fda.gov/media/72033/download.

United States Food and Drug Administration (USFDA), Center for Drug Evaluation and Research (CDER), Center for Biologics Evaluation and Research (CBER). February 2020. "Providing Regulatory Submissions in Electronic Format — Certain Human Pharmaceutical Product Applications and Related Submissions Using the eCTD Specifications Guidance for Industry." https://www.fda.gov/media/135373/download.

United States Food and Drug Administration (USFDA), Office of Pharmaceutical Quality, Center for Drug Evaluation and Research (CDER). November 4, 2015. "Manual of Policies and Procedures." https://www.fda.gov/media/94381/download.

4 Basics of Generic Drug Regulation

Giridhar Prasad and Anurag Patel

HISTORY

The Food and Drug Administration (FDA) is part of the Department of Health and Human Services (DHHS). Among other roles, the FDA is responsible for protecting public health by assuring the safety and efficacy of drugs and biologics intended for human use. The FDA is also responsible for advancing public health by helping to speed innovation.[1]

Over the decades, the FDA has gradually strengthened its regulatory expectations for marketed drugs, starting with the Food and Drugs Act of 1906,[2] followed by the Food, Drug and Cosmetic Act of 1938,[3] and most recently, the Kefauver-Harris Amendments of 1962.[4] The Kefauver-Harris Amendments were enacted in response to the thalidomide tragedy that resulted in thousands of birth defects in newborn children. This amendment mandates the pre-market approval of drug products to include both safety and efficacy, oversee prescription drug advertising, and establish and enforce good manufacturing practices. This amendment also required the FDA to assess the efficacy of all drugs introduced since 1938. The FDA, in response to the new requirements, initiated a process to study drugs approved for safety only between 1938 and 1962 relative to the new requirement for demonstrating efficacy, through a contract with the National Academy of Sciences' National Research Council in 1966. The Drug Efficacy Study Implementation (DESI) program evaluated more than 3,000 separate products and 16,000 therapeutic claims.[5]

Safety and efficacy are of paramount importance for any drug product and are established by the innovator drug through preclinical tests performed on laboratory animals, followed by clinical studies in humans in a phased manner. The safety margin of a drug is between the highest tolerable dose that causes severe or life-threatening side effects and the optimal dose needed to achieve the desired benefit. Studies that evaluate safety also seek to identify any potential adverse effects that may result from exposure to the drug. Efficacy refers to whether a drug demonstrates a health benefit over a placebo or other intervention when tested in an ideal situation, such as a tightly controlled clinical trial. Effectiveness describes how the drug works in a real-world situation. Effectiveness is often lower than efficacy because of, for example, interactions with other medications or health conditions of the patient, or if a sufficient dose or duration of use was not prescribed by the physician or followed

DOI: 10.1201/9781315117911-4

by the patient. These studies are time-consuming, expensive, and prohibitive to lowering the cost of drugs.

To address the high prices of drugs, the *Drug Price Competition and Patent Restoration Act* of 1984, also known as the *Hatch-Waxman Act,* was enacted by Congress mainly due to economic pressures from the growing US elderly population and an increasingly expensive health care system. The high cost of drug prescriptions was blamed on the lengthy pre-approval process for new drug products requiring full safety and efficacy data, which consequently resulted in a costly and lengthy drug development cycle. With the passage of this Act, the FDA has the authority to regulate generic drugs using a new paradigm that ensures they are safe and effective. This paradigm allows for the pre-approval of generic drugs based on a scientifically sound regulatory pathway while managing the impact on new drugs and their patents.

The Hatch-Waxman Act was created in response to a federal district court case involving the drug flurazepam, Roche Products, Inc. v. Bolar Pharmaceutical Co., 733 F.2d 858 (Fed. Cir. 1984).[6] Roche marketed flurazepam, and Bolar was seeking to market a copy of the drug after Roche's patents expired. Bolar was hoping to time the approval with the expiration of the Roche patents and began work on drug development activities while they were still in force. Roche sued Bolar for patent infringement and ultimately won in court. This case set a precedent that restricted another manufacturer from developing a drug while its patent was still in force. Due to the lengthy drug development process of about two to three years, this essentially has the effect of extending the patent life. Congress acted quickly after Roche Products, Inc. v. Bolar Pharmaceutical Co. was decided to pass the Hatch-Waxman Act.

The Hatch-Waxman Act provides incentives for generic drug companies to challenge patents held by innovators, and gives generics a research exemption that allows them to develop generic drugs while patents for the brand are still in force, without liability of infringement. This allows the generic to enter the market as quickly as possible upon patent expiration of the innovator drug. Hatch-Waxman also has provisions that are favorable to innovators. Innovator drugs enjoy a period of FDA exclusivity, normally five years past their patent expiry, during which the FDA does not approve a generic. In addition, Hatch-Waxman gives innovators a potential five-year patent term extension. For several big drugs, these periods of exclusivity have been worth billions of dollars.

Hatch-Waxman added the Abbreviated New Drug Application (ANDA) pathway (505(j) to the FD&C Act) for approval of a generic drug using the same active ingredient as the innovator product. This approval pathway is appropriate for products for which there is therapeutic equivalence. Amendments to the Hatch-Waxman Act resulted in the FDA publishing the Orange Book[6a] (reference Orange Book here). The Orange Book and its monthly Cumulative Supplements make publicly available a list of approved drug products along with their therapeutically equivalent generic versions. The Addendum to this publication identifies drugs that have qualified under the FD&C Act for periods of exclusivity and provides patent information concerning the approved drug products in the Orange Book, among other aspects that may be useful to generic manufacturers and the public.

Generic drugs are not defined in the Federal Food, Drug, and Cosmetic Act of 1938 (FD&C Act) or in US Food and Drug Administration (FDA) regulations. The term "generic drugs" is generally used to describe drug products that have the same active ingredient, dosage form, strength, and route of administration as a "brand-name" drug, and for which FDA has concluded that the generic can be substituted for the brand-name drug. A generic drug prescription is intended to work in the same way and provide the same clinical benefit as the "brand-name" medicine. In other words, you can take a generic drug as an equal substitute for its brand-name counterpart. Currently, generic drugs account for about 90% of the prescriptions filled in the United States.[7]

GENERIC DRUGS OVERVIEW

The modern era of generic drugs was born with the passage of the Hatch-Waxman Act. This act expedited the availability of less-costly generic drugs by permitting the FDA to approve abbreviated applications after the innovator drug's patent had expired. These are termed "abbreviated" applications because they are not required to include animal safety and clinical data to establish safety and effectiveness.

The approval process of generic drugs does NOT require the submission of clinical data to support safety and efficacy since this information will already have been provided by the RLD product. However, a generic drug must be therapeutically equivalent to the branded product. The generic product must meet the same requirements for identity, strength, purity, and quality as the innovator drug. In addition, for the generic drug to be therapeutically equivalent, two clinical characteristics must apply, which are that the generic drug should be a) pharmaceutically equivalent and b) bioequivalent. Pharmaceutical equivalence requires that the active ingredient(s), dosage form, route of administration, and strength are the same for both the generic and innovator drug products, although they do not need to contain the same inactive ingredients (for most dosage forms). Bioequivalence is when both products have comparable bioavailability (i.e., rate and extent of absorption) when studied under similar conditions.

The abbreviated new drug application (ANDA) for submission to the FDA under section 505(j) of the Federal Food, Drug, and Cosmetic Act (FD&C Act) (21 U.S.C. 355(j)) must contain the following information:

- Bioequivalence data
- Quality data
- Administrative information (including labeling and patent certification information)

During the review of the generic drug product application, FDA will evaluate the data presented in these three components to ensure the safety and efficacy of the generic product. Figure 4.1

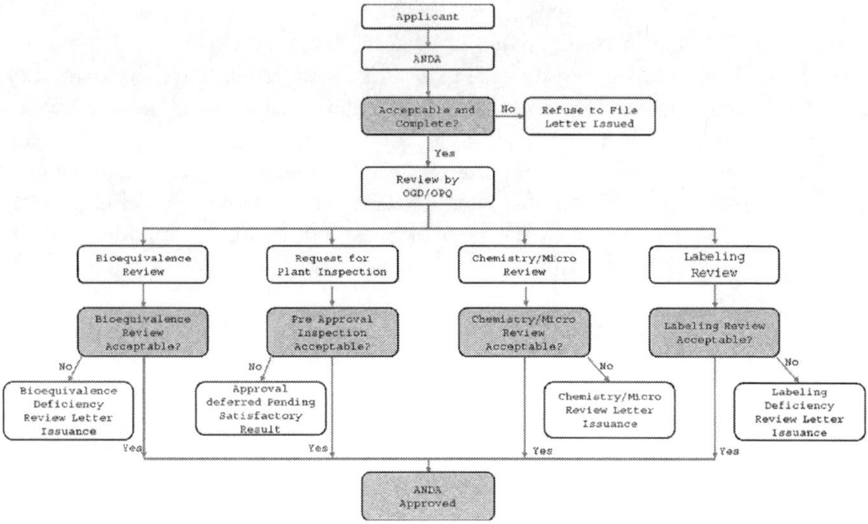

FIGURE 4.1 ANDA approval process

Bio Equivalence Review Process

A generic drug is considered bioequivalent to RLD if

- The rate and extent of the drug's absorption do not show a significant difference from the RLD's rate and extent of absorption when administered at the same molar dose of the therapeutic ingredient under similar experimental conditions in either a single dose or multiple doses at the site of drug action; or
- The extent of the drug's absorption does not show a significant difference from the extent of the RLD's absorption when administered at the same molar dose of the therapeutic ingredient under similar experimental conditions in either a single dose or multiple doses, and the difference from the listed drug in the rate of absorption of the drug is intentional, reflected in its proposed labeling, not essential to the attainment of effective body drug concentrations on chronic use, and considered medically insignificant for the drug. (FD&C Act Section 505(j)(8)(B); 21 CFR 320.1(e))

Bioequivalence is determined by the evaluation of the AUC and the maximum concentration of the drug (Cmax). A generic product is bioequivalent to the branded product if the 90% confidence interval (CI) of the mean AUC and the relative mean Cmax is 80% to 125%. In a standard in vivo bioequivalence clinical study, single doses of the test and reference drug products are administered to volunteers, and the rate and extent of the drug's absorption are determined from measured plasma concentrations over time for each subject participating in the study. The extent of absorption (i.e., how much of the drug in the given dose was absorbed) is reflected through various measurements. The preferred study design for oral dosage forms is a single-dose, fasting, two-treatment, two-sequence crossover design with a washout period between treatments.[8] In a crossover study, the subjects receive the test and reference drug products in separate sequences (either test before reference or reference before test), with a period between treatments of no drug administration (the washout period) to ensure the previous dose is cleared from the body before the second dose is administered. To demonstrate in vitro bioequivalence, the FDA recommends an applicant use an in vitro test (e.g., a dissolution rate test) correlating with in vivo data.[9] Over the past few years, the FDA has begun to publish individual product bioequivalence recommendations on its website. For more details, refer to https://www.ncbi.nlm.nih.gov/pmc/articles/PMC4157693/.

Applicants may also request a waiver from performing in vivo (testing done in humans) bioequivalence studies for certain drug products (21 USC 360 b (n) (1) (E)) where bioavailability may be demonstrated by submitting data such as 1) a formulation comparison for products whose bioavailability is self-evident, for example, oral solutions, injectables, or ophthalmic solutions where the formulations are identical (Q1, Q2) or 2) comparative dissolution. These biowaiver applicabilities for the generic formulation can be requested during the generic product development phase through controlled correspondence.[10]

In general, the generic product being considered for a biowaiver must contain the same active and inactive ingredients (Q1) in the same dosage form and concentration (Q2), and some formulations may be required to have the same pH and physicochemical characteristics (Q3) as an approved innovator product. These are further evaluated during the quality review.[11]

Quality Review Process

The aim of this review, sometimes referred to as a CMC review, is to provide assurance that the generic drug will be manufactured in a reproducible manner under controlled conditions. Areas such as the applicant's manufacturing procedures, raw material specifications, manufacturing controls, sterilization process, container closure systems, and stability data are reviewed to assure that they are adequate for assuring and preserving the identity, strength, quality, and purity of the proposed drug product. The CMC review is consistent with the rigor applied to the innovator drug and includes an evaluation of the pharmaceutical equivalence of the generic to the innovator drug.

For the evaluation of the ANDA application, the Office of Generic Drugs (OGD) has developed the concept of Question-based Review (QbR), which mainly focuses on the critical quality attributes. The Question-based Review has transformed the CMC review into a modern, science- and risk-based pharma-critical quality assessment. The objectives of this Question-based Review system include[12]

- Assuring the product quality through design and performance-based specifications

- Through risk assessment, facilitating continuous improvement and reducing CMC supplements
- Setting standardized review questions to enhance the quality of reviews
- Reducing the CMC review time

The Quality section of the ANDA consists of two sub-classifications. One is the quality information pertaining to the drug substance (API), and the other is the quality information pertaining to the drug product. Both these sections are reviewed in conjunction, and the final approval for the ANDA is dependent upon the successful completion of the review of both quality sections. Reviewers of the drug substance and drug product from the Office of Pharmaceutical Sciences are provided with the QbR templates for the ANDA evaluation. (More details can be found in the ICH M4Q guidance – https://database.ich.org/sites/default/files/M4Q_R1_Guideline.pdf.)

INACTIVE INGREDIENTS

As described above, pharmaceutical equivalence of the generic product to the RLD is required. Pharmaceutical equivalents are drug products using the identical active drug ingredient, i.e., the same salt or ester of the same therapeutic moiety that is formulated in the same dosage form to be delivered through the same route of delivery. However, pharmaceutical equivalents are not necessarily required to contain the same inactive ingredients.

As per the FD&C Act Section 505(j)(4)(H), the FDA cannot approve the ANDA unless, among other things:

(i) The inactive ingredients of the drug are unsafe for use under the conditions prescribed, recommended, or suggested in the labeling proposed for the drug; or
(ii) The composition of the drug is unsafe under such conditions because of the type or quantity of inactive ingredients included or the way the inactive ingredients are included.

There are some exceptions or changes that are allowed in the inactive ingredient for certain types of generic drug products which are set forth in 21 CFR 314.94(a)(9). For example, FDA's regulations for parenteral drug products at 21 CFR 314.94(a)(9)(iii) state that a drug product intended for parenteral use shall contain the same inactive ingredients and in the same concentration as the RLD identified by the applicant under paragraph (a)(3) of this section. Most generic products approved under these requirements also qualify for a biowaiver (see section above). However, approval for a generic drug for parenteral products can be sought if the generic drug is different from the RLD in the use of a preservative, buffer, or antioxidant, provided that the applicant identifies and characterizes the differences and provides information demonstrating that the differences do not affect the safety or efficacy of the proposed drug product.

Preservative, buffer, and antioxidant changes in generic parenteral drug products are referred to as "exception excipients," which may differ qualitatively or quantitatively from the RLD formulation. Other regulations under 21 CFR 314.94(a)(9)(iv) identify exception excipients for generic ophthalmic and otic drug products (i.e., preservative, buffer, substance to adjust tonicity, and thickening agent). Excipients not identified in these regulations are referred to as "non-exception excipients."

The FDA's exception excipient regulations under 21 CFR 314.94(a)(9) find their parallel in 21 CFR 314.127(a)(8)(ii), which addresses the grounds for an FDA refusal to approve an ANDA for a parenteral, ophthalmic, or otic drug product. For example, one regulation states:

> FDA will consider an inactive ingredient in, or the composition of, a drug product intended for parenteral use to be unsafe and will refuse to approve the [ANDA] unless it contains the same inactive ingredients, other than preservatives, buffers, and antioxidants, in the same concentration as the listed drug.[13]

Thus, it is important for the applicant to understand what changes are permitted in the drug product formulation. If a change is for a non-exception excipient, the FDA would otherwise refuse to receive or approve an ANDA. To avoid such a situation, the applicant would need to submit a 505(b)(2) application for the proposed drug product.

INACTIVE INGREDIENTS DATABASE (IID)

As described above, inactive ingredients used in the generic drug are not required to be the same as the RLD for most dosage forms. Nevertheless, inactive ingredients play a vital role in generic drug development. Using the same or similar inactive ingredients as the RLD can, in general, be a reasonable strategy as the starting point for some generic products using unique excipients. However, inactive ingredients also have a safety profile which can affect the application filing and scientific review. To help the industry with the selection of their inactive ingredients for a given dosage form and route of administration, the FDA published the Inactive Ingredients Database (IID).[14] The IID confirmations can be requested during the development phase through controlled correspondence.[10] The IID has been used by both innovator and generic manufacturers as an aid in developing their drug products. If an excipient is used in approved drug products, the excipient generally is not considered new and may warrant less extensive review the next time it is included in a new drug product.[15] For example, if the IID lists an excipient at a specified potency for a certain route of administration and dosage form, a generic manufacturer could generally consider the excipient at that potency safe for use in a similar manner for a similar type of product. The IID, however, has limitations regarding the different exposure models that must be justified or supported with evidence of supporting safety data during the substantial review of the application. These limitations include support for the maximum daily intake based on the dosing recommendations indicated in the labeling, safety in pediatric populations, and acute versus chronic use.

Administrative Information Review Process

Labeling Review

The labeling review process ensures that the proposed generic drug labeling (package insert, container, package label, and patient information) is identical to that of the reference listed drug, except for differences due to changes in the manufacturer, distributor, pending exclusivity issues, or other characteristics inherent to the generic drug product (excipients, tablet size, shape, or color, etc.). Furthermore, the labeling review serves to identify and resolve issues that may contribute to medication errors, such as similar sounding or appearing drug names as applicable where proprietary names are requested, and the legibility or prominence of the name or strength.

After the final level administrative review and individual disciplines ascertained that all the information received is satisfactory the application will either receive a full approval or a tentative approval letter. A full approval letter details the conditions of approval and allows the applicant to market the generic drug product. A tentative approval letter may be issued if there are unexpired patents or exclusivities accorded to the Reference Listed Drug (RLD).

THERAPEUTIC EQUIVALENCE

So far, we have discussed how the ANDA review process works and what information is required for the ANDA to be approved. Provided that all the aspects of the ANDA application are met, the application is now ready for approval. However, before the approval can happen, the FDA must decide whether the subject ANDA product is therapeutically equivalent to the reference listed product. In this section of the chapter, we will discuss therapeutic equivalence and how the products are rated.[16]

The term "therapeutic equivalence," which is quite commonly used for generic products, is not defined in the FD&C Act or in the FDA's ANDA regulations. However, "drug products are considered to be therapeutic equivalents only if they are pharmaceutical equivalents and if they can be expected to have the same clinical effect and safety profile when administered to patients under the conditions specified in the labeling" (i.e., are bioequivalent). "FDA believes that products classified as therapeutically equivalent can be substituted with the full expectation that the substituted product (Generic Product) will produce the same clinical effect and safety profile as the reference product."

Alternatively, drugs that are not therapeutic equivalents and are not substitutable for the RLD relied upon for approval are termed "pharmaceutical alternatives." The FDA's regulations define "pharmaceutical alternatives" as:

Drug products that contain the identical therapeutic moiety, or its precursor, but not necessarily in the same amount or dosage form or as the same salt or ester. Each such drug product individually meets either the identical or its own respective compendial or other applicable standard of identity, strength, quality, and purity, including potency and, where applicable, content uniformity, disintegration times, and/or dissolution rates.[17] Thus, products approved pursuant to an approved suitability petition (i.e., a change in strength, route of administration, dosage form, or active ingredient in a combination drug) or 505(b)(2) are pharmaceutical alternatives and are not listed in the Orange Book as therapeutically equivalent to the RLD.

Pharmaceutically equivalent prescription drug products usually are identified in the Orange Book with either an "A" or "B" therapeutic equivalence code designation, which is further defined by a one-character subcode.[18]

"A-rated" drug products are considered to be therapeutically equivalent to other pharmaceutically equivalent products because there are no known or suspected bioequivalence problems, or such problems have been resolved with adequate evidence supporting bioequivalence. These products fall under one of the categories below:

- Active ingredients or dosage forms for which no in vivo bioequivalence issue is known or suspected, and for which bioequivalence to the RLD is presumed and considered self-evident based on other data in an application or by showing an acceptable in vitro dissolution standard is met.
- Active ingredients or dosage forms presenting a potential bioequivalence problem, but the applicant's approved application contains adequate scientific evidence establishing (through in vivo and/or in vitro studies) bioequivalence of the product to a selected RLD.

Drug products falling under the first category are assigned a therapeutic equivalence code depending on the dosage form. These codes are AA, AN, AO, AP, or AT. Drug products falling under the second category are coded AB, the most common code assignment.

"B-rated" drug products are not considered to be therapeutically equivalent to other pharmaceutically equivalent drug products because actual or potential bioequivalence problems identified by the FDA have not been resolved by adequate bioequivalence evidence. These products fall under one of the categories below:

- The FDA has documented bioequivalence problems or a significant potential for such problems (and no adequate studies demonstrating bioequivalence have been submitted to the Agency).
- The FDA has an insufficient basis to determine therapeutic equivalence (e.g., quality standards are inadequate).
- The drug products are under regulatory review.

Drug products falling under these categories are assigned a therapeutic equivalence code largely based on dosage form. These codes are B, BC, BD, BE, BN, BP, BR, BS, BT, or BX.

PATENT AND EXCLUSIVITIES

As discussed in this chapter, an ANDA applicant relies entirely upon the safety and efficacy data generated by the Reference Listed Drug (RLD), and ANDA approval is solely based on the sameness to the RLD. The Hatch-Waxman Act also puts mechanisms into place to protect the RLD's patent rights from companies relying on the RLD's clinical data while allowing drug developers (innovators and generics alike) to conduct the work necessary to secure approval without infringing the RLD owner's patents, and allowing generic drugs to be marketed as early as possible after an RLD's patents have expired.

An ANDA for a generic version of an RLD, amongst other things, must also contain one of four possible certifications "with respect to each patent which claims the reference drug or which claims a use, for such listed drug and for which information is required to be filed" by the NDA holder and is listed in the Orange Book (FD&C Act Section 505(j)(2)(A)(vii)). The four different types of certification that are required to be part of the ANDA are as follows:

Paragraph (I): That no patent information on that brand name drug has been submitted to the FDA.

Paragraph (II): That the listed patent has expired.

Paragraph (III): That the listed patent will expire on a certain date, before which time the generic will not enter the market.

Paragraph (IV): That the patent is invalid or will not be infringed by the manufacture, use, or sale of the new drug for which the ANDA was submitted.

When the generic company submits a paragraph I or II certification, the FDA may approve its ANDA immediately. The FDA may approve a paragraph III certification any time after the patent's expiration date. The implications of a paragraph IV certification are not nearly as simple. A generic company submits a paragraph IV certification when it does not want to wait for the expiration of the innovator's patents before it begins to market its own generic version of the drug. Instead, it alleges that it is justified for an early market entry because its drug does not infringe the innovator's patent, or the patent is invalid.

In the event an ANDA application contains the paragraph IV certification, the ANDA applicant is required to send the notice of paragraph IV to the patent holder and the NDA holder, notifying them about such a certification being submitted and explaining the grounds for the notice. As a response to this notice, an NDA or patent owner may choose to sue a generic applicant that submits a paragraph IV certification. If suit is brought, the FDA is statutorily proscribed from approving the ANDA for 30 months, with certain limited exceptions (i.e., the so-called "30-month stay"). If a patent infringement suit is not brought by the NDA or patent owner, the FDA can approve the ANDA once it has determined all conditions required for approval have been met.

ANDA applications also have an alternative to these four certifications if the patent is a "method of use patent which does not claim a use for which the applicant is seeking approval." In this case, the application instead may contain "a statement that the method of use patent does not claim such a use." This often is referred to as a "section viii statement" and permits a generic applicant to "carve out" a patent-protected use from its proposed labeling, provided the omission of such protected information does not "render the proposed drug product less safe or effective than the listed drug for all remaining, non-protected conditions of use."

The FDA has developed various procedural tools to carry out its obligations under the Hatch-Waxman Act, including its use code system for evaluating a section viii statement's propriety. Under this system, the RLD owner submits to the FDA a use code for any method-of-use patents listed in the Orange Book that describes what the patent covers. When evaluating a section viii statement's propriety, FDA will compare the labeling in the ANDA to the use code and determine if the label describes a use claimed in the patent. If the patent covers a use described in the label, the FDA will deny the section viii statement and require the ANDA holder to include an appropriate patent certification.

505(B)(2) NDA – FD&C ACT, SECTION 505(B)(2)

In addition to the 505(j) (ANDA) application pathway, the 505(b)(2) may also be used to obtain pre-market approval to market generic drugs. The 505(b)(2) pathway is suitable for generics when their rate and/or extent of absorption exceed, or are different from, the ANDA standards, and may require studies to show safety and efficacy. The 505(b)(2) applications use a similar active ingredient but introduce a new indication, dosage form, route of administration, strength, combination of two active ingredients, or an Rx to OTC switch, etc.

Similar to the 505(b)(1), the 505(b)(2) pathway was created, in part, to help avoid unnecessary duplication of studies already performed by the innovator of the RLD product and allows the FDA to rely on data not developed by the 505(b)(2) applicant. A 505(b)(2) NDA must contain full safety and effectiveness reports but allows, for example, some required safety and efficacy information for approval to come from studies not conducted by or for the applicant.

A key feature of the 505(b)(2) pathway is that it allows a manufacturer to submit their product for FDA review by referencing the data and/or study results originally collected by another manufacturer or researcher (RLD product). The manufacturer of the 505(b)(2) eligible product needs to build a connection between their version of the product, or the active ingredients in it, and the reference product. For example, these data could be the data and results of bioanalytical testing, pre-clinical studies, or even clinical trial results. If successful in their effort to include supporting evidence from other published literature in their submission, the manufacturer of the 505(b)(2) candidates won't have to re-run these studies themselves. While the 505(b)(2) path allows for using the research of others as a component of their FDA submission, the manufacturer of the 505(b)(2) product may still need to complete some of their own research in other areas to help fulfill all the various requirements of the FDA to earn approval.[19]

A 505(b)(2) type of application can be submitted for

- A New Chemical Entity when some part of the data necessary for approval is derived from studies not conducted by or for the applicant, and to which the applicant has not obtained a right of reference. These types of data are generally derived from published studies rather than the FDA's previous finding of safety and effectiveness of a drug. These can be products that are marketed without approval.
- Seeking approval of a change to a previously approved drug, where an applicant may rely on the FDA's finding of safety and effectiveness of the previously approved product, coupled with the information needed to support the change from the approved product. The additional information could be new studies conducted by the applicant or published data.

  Some of the types of changes that can be eligible for the submission of the 505(b)(2) application are

  Dosage form: An application for a change of dosage form, such as a change from a solid oral dosage form to a transdermal patch, that relies to some extent upon the Agency's finding of safety and/or effectiveness for an approved drug.

  Strength: An application for a change to a lower or higher strength.

  Route of administration: An application for a change in the route of administration, such as a change from an intravenous to an intrathecal route.

  Substitution of an active ingredient in a combination product: An application for a change in one of the active ingredients of an approved combination product for another active ingredient that has or has not been previously approved.

USER FEES

As discussed in this chapter on the various aspects of the generic drug review and approval process, the FDA has come up with a program to speed up the approval of generic drugs and reduce the overall cost to industry. This new program, part of FDASIA legislation, is known as the Generic Drug User Fee Amendments of 2012 (GDUFA). Using the success of the Prescription Drug User Fee Act (PDUFA; the user fee program for NDAs) as its model, the law requires the industry to pay user fees to supplement the costs of reviewing generic drug applications. Two of the key goals are to reduce the backlog of current applications and decrease the overall time required for a decision. Additionally, GDUFA aims to ensure consistency and increase the frequency of inspections for both domestic and foreign facilities. With this added focus on inspections, the FDA hopes to enhance the safety of the global supply chain for generic drugs.

GDUFA

On July 09, 2012, GDUFA was signed into law by the President as part of the Food and Drug Administration Safety and Innovation Act (FDASIA). It is designed to speed up the access to safe and effective generic drugs to the public and reduce their cost to the industry. This act makes it possible for the FDA and industry to continue to ensure that the American public has access to safe and high-quality generic drugs. Under GDUFA, all the firms that manufacture human generic drug products and active ingredients for human generic drug products are subject to FDA user fees. GDUFA enables the FDA to assess industry user fees to bring greater predictability and timeliness to the review of the generic drug application. GDUFA must be reauthorized every five years. GDUFA was reauthorized again on September 30, 2022 (GDUFA III), and remain in effect through September 30, 2027.[20]

REFERENCES

1. https://www.fda.gov/about-fda/fda-basics/fda-fundamentals
2. https://www.fda.gov/about-fda/changes-science-law-and-regulatory-authorities/part-ii-1938-food-drug-cosmetic-act
3. https://www.fda.gov/about-fda/changes-science-law-and-regulatory-authorities/part-ii-1938-food-drug-cosmetic-act
4. https://www.fda.gov/about-fda/changes-science-law-and-regulatory-authorities/part-iii-drugs-and-foods-under-1938-act-and-its-amendments
5. https://www.fda.gov/drugs/enforcement-activities-fda/drug-efficacy-study-implementation-desi
6. https://unctad.org/ippcaselaw/sites/default/files/ippcaselaw/2020-12/Roche%20v.%20Bolar%20Pharmaceutical%20.pdf

6a. Food and Drug Administration. (n.d.). Approved drug products with therapeutic equivalence evaluations (44th ed.). U.S. Department of Health and Human Services. https://www.accessdata.fda.gov/scripts/cder/ob/

7. https://www.fda.gov/drugs/buying-using-medicine-safely/generic-drugs
8. https://www.fda.gov/files/drugs/published/Guidance-for-Industry-Bioavailability-and-Bioequivalence-Studies-for-Orally-Administered-Drug-Products---General-Considerations.PDF
9. 21 CFR 320.24(b)(5)
10. https://www.fda.gov/regulatory-information/search-fda-guidance-documents/controlled-correspondence-related-generic-drug-development-guidance-industry
11. https://www.fda.gov/animal-veterinary/abbreviated-new-animal-drug-applications/bioequivalence
12. Question-Based Review (QbR) for Generic Drugs: An Enhanced Pharmaceutical Quality Assessment System. Center for Drug Evaluation and Research. 2018. [cited 2019 Feb 6]. https://www.fda.gov/Drugs/DevelopmentApprovalProcess/HowDrugsareDevelopedandApproved/ ApprovalApplications/AbbreviatedNewDrugApplication ANDAGenerics/ ucm120973.htm
13. 21 CFR 314.127(a)(8)(ii)(B)
14. https://www.fda.gov/drugs/drug-approvals-and-databases/inactive-ingredients-approved-drug-products-search-frequently-asked-questions

15. https://www.fda.gov/media/128687/download
16. https://www.fda.gov/drugs/development-approval-process-drugs/orange-book-preface
17. FD&C Act Section 505(j)(8)(B); 21 CFR 320.1(e)
18. 21 CFR 320.1(d)
19. https://www.fda.gov/media/72419/download
20. U.S. Food and Drug Administration. (2022). Generic Drug User Fee Amendments (GDUFA III) Reauthorization. Retrieved from https://www.fda.gov/industry/fda-user-fee-programs/generic-drug-user-fee-amendments

5 Introduction to Medical Device Regulation

Anurag Patel

HISTORY OF DEVICES

Medical devices have been used by ancient civilizations throughout history. Indeed, surgical instruments and treatments (CE 1600 BCE) describing their use have been discovered at excavation sites in Egypt (Sullivan 1996). The art of plastic surgery was practiced in India as early as 2000 BCE using surgical instruments. The Sushruta Samhita (Susruta and Bhishagrantna 1907–1916), a medical instruction manual, describes and states the intended use of about 120 surgical instruments for over 300 surgical procedures. Surgical devices were also used by the Romans and in medieval times across the West and brought to the New World. Medical devices were marketed without regulatory controls in most of the world, including the US. In the US, the first device to be exposed as fraudulent was in 1799 (Rados 2006). Dr. Elisha Perkins sold a set of brass and iron rods, fraudulently claiming that their use could eliminate disease from the body. Even when enacting the Food and Drug Regulations of 1906, devices of the time, such as stethoscopes and scalpels, were seen as comparatively simple, and any hazards or defects were determined to be readily apparent. Devices, therefore, were not perceived to pose a significant threat and were excluded from the new law. The Agency reacted by stating in its annual report to Congress that the 1906 act "has its serious limitations ... which render it difficult to control ... fraudulent mechanical devices used for therapeutic purposes." In 1898, French chemists Pierre and Marie Curie discovered the radioactive element radium. This discovery added even more fraudulent devices with exaggerated claims attributed to their radioactive properties. By 1917, the FDA began observing a flood of fraudulent devices marketed to the US public, such as nose straighteners, height-stretching machines, and heated rubber applicators advertised as a cure for prostate gland disorders. Another report to Congress in 1926 continued to bring to its attention the exaggerated health claims and use of products containing radium. The Agency, having no direct authority over medical devices, continued to monitor the products and enforced the removal of fraudulent devices from the market with assistance from the Federal Trade Commission and the US Post Office. The Federal Trade Commission and the US Post Office were both assigned to oversee devices and enforce criminal penalties for mail fraud under the Postal Fraud Statutes. In 1938, the enactment of the Federal Food, Drug, and Cosmetic Act (FD&C Act) included

DOI: 10.1201/9781315117911-5

the definition of a medical device and allowed the Agency to bring federal charges in the courts against therapeutic devices or materials that were found to be defective, unsafe, filthy, or otherwise produced in unsanitary conditions (adulterated), or against statements, designs, or labeling that were false or misleading (misbranded). There was, however, no requirement for pre-market testing, review, or approval in this law. After World War II, the number of marketed devices skyrocketed due to inventors having access to large amounts of electronic parts such as dials, knobs, and meters from military surplus. A greater proportion of devices marketed during this period had no medical benefit, were making fraudulent claims, and in some cases were also harmful. The RelaxiAcizer was one such device, first invented in 1938 (Suzanne 2018). The electrical machine applied periodic electrical impulses, through electrodes placed at various parts of the body, to induce muscle contraction and relaxation cycles. This device was heavily marketed as a passive exercise machine for effortless weight loss and gained popularity between 1960 and 1970. A lawsuit was eventually filed by the FDA to remove the RelaxiAcizer from the market. After five months of litigation, in April of 1970, the device was finally ruled by the US district court in LA to be dangerous, and a permanent injunction was issued against its sale. The FDA continued a public warning campaign against other devices similar to the RelaxiAcizer that continued to be on the market.

In 1962, President John F. Kennedy proposed changing the regulation of medical devices as they were introduced into the market. As a result, Congress heard proposals to revise the FD&C Act in a number of different ways, including a requirement that medical devices be regulated comparably to, but separately from, new drugs (Rados 2006). At the same time, in response to the news of the thalidomide tragedy in Europe, Congress enacted the 1962 Kefauver Harris Amendment, or "Drug Efficacy Amendment." However, a companion bill to require premarket approval for devices was again not enacted into law.

During the 1960s and into the 1970s, the Agency continued monitoring fraudulent medical devices already on the market but also needed to manage a wide range of new life-saving medical devices as they entered the market. This strained the Agency's resources to the limit. President Richard M. Nixon's endorsement of medical device legislation in 1970 initiated the Cooper Committee, which suggested creating different classifications for medical devices to tailor regulatory controls against the risks involved. As these recommendations were being debated in Congress during 1972 and 1973, pacemaker malfunctions leading to device failures were reported. This was followed by the 1975 congressional hearings reporting problems with the Dalkon Shield intrauterine device, which caused thousands of reported injuries. These two incidents helped underscore the need for the Medical Device Amendments, enacted in 1976 (94th US Congress 1976). Due to congressional indecision, the Agency had already taken the recommendations from the Cooper Committee. The Agency prepared a medical device inventory and classified these devices into three groups based on risk. The lowest risk Class I devices would fall under general controls, higher risk Class II devices would be subjected to performance standards when general controls were deemed insufficient to assure product safety and effectiveness, and the highest risk Class III devices would be required to go through pre-market approval. The

Medical Device Amendments of 1976 additionally enacted new cGMP requirements for medical devices and required Class III device manufacturers to notify the FDA prior to their marketing. New medical devices that were substantially equivalent to pre-1976 devices could be marketed immediately, subject to any existing or future requirements for that type of medical device. In spite of its significant enhancement to device regulations, the Medical Device Amendments of 1976 lacked adequate provisions to protect the public against dangerous medical devices already on the market. The Safe Medical Devices Act (SMDA), passed in 1990, provides the Agency with enforcement authority to order product recalls and other actions. This law requires medical device manufacturers and various health care facilities that use medical devices to report device malfunctions to the Agency. Manufacturers of permanently implanted devices whose failure might cause serious harm or death are required to conduct post-market surveillance and to establish methods for tracking and tracing such devices. The medical device cGMP requirements or Quality System Requirements (QSR) were further strengthened on December 18, 1978, in 21 CFR 820, as authorized by Congress under section 520 (f) the FD&C Act. The Agency notes in the QSR final rule that 90 percent of software-related recalls between 1983 and 1991 were due to design-related errors and the failure to validate software prior to routine production. The reader is referred to an in-depth discussion on the medical device QSR later in this chapter.

More recently, the Center for Devices and Radiological Health (CDRH) within the Agency has responded to significant market disruptors, including smartphones, social media, and the sharing economy. Many of the technological leaps causing these market disruptions originate from complex digital technologies that introduce new functionalities, such as interconnectivity, interoperability, cybersecurity, and artificial intelligence. The technology, along with its pace of development and change, introduces new risks to the public and requires a new paradigm for regulatory governance. CDRH has taken a proactive approach to these challenges by hiring staff and enhancing its digital health resources through collaborations at all levels: interdepartmental, intradepartmental, and external (Food and Drug Administration n.d.).

The Food and Drug Administration Safety and Innovation Act (FDASIA) (Pub. L. 112-144) enacted on July 9, 2012, required the FDA, in consultation with the Office of the National Coordinator for Health Information Technology (ONC) and the Federal Communications Commission (FCC), to propose and provide a report (The Office of the National Coordinator for Health Information Technology 2014) of its strategy and recommendations for an appropriate, risk-based regulatory framework pertaining to functionalities such as health information technology (IT), including mobile medical applications, that promotes innovation, protects patient safety, and avoids regulatory duplication. The FDA would specifically focus its oversight to provide assurance of safety and effectiveness for these functionalities on their medical device functionality risks to patient safety. The workgroup concluded the need for collaborative interagency requirements with FCC, the distinction between wellness and disease-related medical device claims, and the definition of medical device accessories, clinical decision support software, and medical device software modules.

As a consequence, the FDASIA section 520 of the FD&C Act was amended on December 13, 2016, through the 21st Century Cures Act (Cures Act). This amendment removes certain software functions from the definition of device in section 201(h) of the FD&C Act. CDRH published the Digital Health Innovation Action Plan (Food and Drug Administration n.d.) to set forth what it saw as the next steps to publish guidance and policy documents to communicate its interpretation of the medical software provisions in the Cures Act (Food and Drug Administration 2019). The FDA also communicated its intent to develop a precertification program that could replace the need for a premarket submission for certain products and allow for decreased submission content and/or faster review of the marketing submission for other products (Food and Drug Administration n.d.). The Agency continues to iterate and build on its test program (Food and Drug Administration n.d.).

Center for Devices and Radiological Health (CDRH)

The FDA's Center for Devices and Radiological Health (CDRH) is responsible for regulating medical devices and radiation-emitting electronic products intended for both medical and non-medical purposes in the United States market. CDRH regulates products with a wide variety of technologies and intended uses, e.g., medical devices such as manual toothbrushes, bedpans, programmable pacemakers, digital devices, and closed-loop artificial pancreas systems; in vitro diagnostic (IVD) products such as reagents, test kits, and blood glucose meters; and ionizing and non-ionizing radiation-emitting electronic products (medical and non-medical) such as lasers, x-ray systems, ultrasound equipment, microwave ovens, and color televisions.

The CDRH is one of the nine center-level organizations within the Department of Health and Human Services. The CDRH consists of six offices that are organized under the Office of the Center Director, of which the Office of Product Evaluation and Quality is a super office. This office consists of another ten offices. The remaining five offices may be further broken down to the division level. The CDRH mission, vision, and organizational chart, along with functions and responsibilities of its various offices and divisions, are published on its website. The reader is referred to the FDA website for the most current information with more details.

REGULATION AND CLASSIFICATION OF MEDICAL DEVICES

There is a wide variety of medical devices incorporating an equally broad range of technologies. Medical devices may be as simple as toothbrushes and bedpans or include more technologically advanced devices like programmable pacemakers and artificial intelligence software. A subset of medical devices that are used to test blood or tissue samples for diagnostic purposes are further classified as in vitro medical devices, e.g., reagents, test kits, and blood glucose meters. The FDA defines a medical device in the Food, Drug, and Cosmetic Act (21 USC 321(h)).

USC 321(H)

(1) *The term "device" ... means an instrument, apparatus, implement, machine, contrivance, implant, in vitro reagent, or other similar or related article, including any component, part, or accessory, which is –*

(A) *recognized in the official National Formulary, or the United States Pharmacopeia, or any supplement to them,*

(B) *intended for use in the diagnosis of disease or other conditions, or in the cure, mitigation, treatment, or prevention of disease, in man or other animals, or*

(C) *intended to affect the structure or any function of the body of man or other animals, and*

which does not achieve its primary intended purposes through chemical action within or on the body of man or other animals and which is not dependent upon being metabolized for the achievement of its primary intended purposes. The term "device" does not include software functions excluded pursuant to section 360j(o) of this title.

The distinction between a "general" device and a medical device is made based on its intended use and indications of use (Table 5.1). A pocketknife and surgical scalpel are both devices that function through a physical action, as cutting tools. The intended use of the surgical scalpel classifies it as a medical device.

IN VITRO MEDICAL DEVICES

In vitro diagnostic products are defined in 21 CFR Part 809. In vitro diagnostic products are a subset of medical devices. These devices may include reagents, instruments, and systems intended for use in the diagnosis of disease or other conditions, including a determination of the state of health, in order to cure, mitigate, treat, or prevent disease or its sequelae. In vitro medical devices may also be intended for use in the collection, preparation, and examination of specimens taken from the human body. In vitro medical devices have specific requirements for labeling in accordance with 21 CFR 809.10.

TABLE 5.1
Criteria Defining Medical Devices

Intended Use	Indications for Use
The general purpose of the device or its function. This includes the indications for use.	Describes the disease or condition the device will diagnose, treat, prevent, cure, or mitigate, including a description of the patient population for which the device is intended.

TABLE 5.2
Regulatory Requirements Based on Device Classification and Risk

Class	Risk	Potential Harm	Regulatory Controls	Submission Type or Exemption
I	Lowest	Minimal potential for harm e.g., oral medicinal cup, tongue depressors	General Controls	510(k) 510(k) exempt
II	Moderate	Lower risk than Class III devices e.g., surgical gowns and surgical masks	General and Special Controls (if available)	510(k) 510(k) exempt
III	Highest	Highest risk e.g., devices that sustain or support life, are implanted, or present potential unreasonable risk of illness or injury	General and PMA	Premarket Approval (PMA)

Classification of Devices

As explained in the previous section, the FDA Medical Device Amendments, enacted in 1976, categorizes medical devices into three classes (I, II, and III), based on the degree of risk they present. FDA classifies medical devices intended for human use in 21 US Code § 360c. As device class increases from I to III, the degree of risk presented by the device and its regulatory controls also increase (Table 5.2).

Further, as device class increases from Class I to Class II to Class III, the regulatory controls also increase, with Class I devices subject to the least regulatory control and Class III devices subject to the most stringent regulatory control. The classes of devices, regulatory controls, and submission types are summarized in Table 5.2.

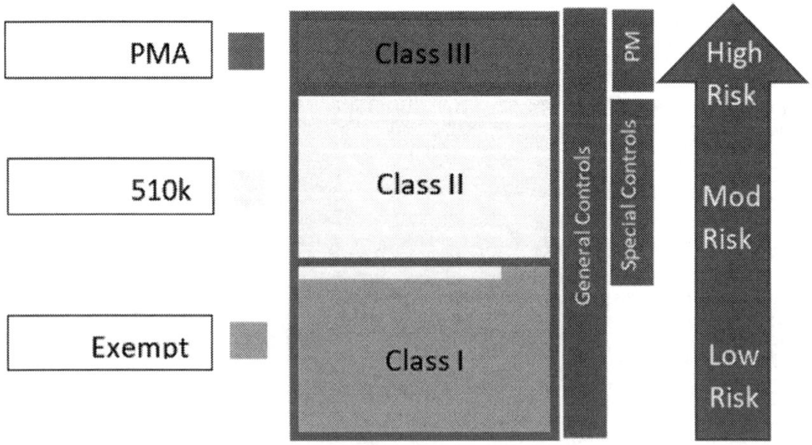

Many, but not all, Class I and some Class II devices are exempt from submission requirements and do not require market clearance or premarket notification from the FDA. Medical device classification may be determined by identifying the device of interest in a list of over 1,700 distinct types of devices classified by the FDA. This list is organized into 16 medical specialty groupings between 21 CFR parts 862 to 892. The individual device listing in the CFR consists of a brief description, intended use and indication of use, device classification, applicable exemptions, and information about marketing requirements. Certain low-risk Class I devices are exempt from general controls, such as GMP requirements, but not from keeping complaint files (21 CFR 820.198) or from general requirements concerning records (21 CFR 820.180).

All Class I and II exempt medical devices are subject to certain limitations. These limitations are described under 21 CFR xxx.9, where xxx refers to Parts 862–892. Devices listed in the CFR may also be searched in the FDA Product Code Classification Database. A formal device determination or classification from the FDA may be obtained by submitting a 513(g) request.

General controls such as registration and listing, labeling, and good manufacturing practices apply to all classes of medical devices. Only a small number of low-risk Class I devices are exempt from GMPs, with the exception of general recordkeeping requirements and complaint files.

Some Class I and most Class II devices are not exempt and require a 510(k) submission before placing them on the market (Food and Drug Administration 1998). The 510(k) submission process (which is discussed later in this chapter) for a new device is also referred to as a premarket notification or 510(k) clearance. For some Class II devices, general controls alone are insufficient to provide reasonable assurance of the safety and effectiveness of the device. For these devices, the FDA additionally requires device-specific special controls to provide such assurance. Special controls may include performance standards, postmarket surveillance, patient registries, special labeling requirements, premarket data requirements, and requirements described in guidelines.

Class III devices are subject to the approval of a Premarket Approval Application (PMA) before placing them on the market (21 CFR 814). Class III devices are classified with the highest risk based on their intended use to support or sustain human life, prevent impairment of human health, or present a potential unreasonable risk of illness or injury. General controls and special controls are insufficient to provide reasonable assurance of the safety and effectiveness of these devices, or for which there is insufficient information to make such a determination. Approval of the PMA is the FDA process of scientific and regulatory review to assure that the device is safe and effective for its intended use(s).

MEDICAL DEVICE ACCESSORIES

An accessory is intended to support, supplement, and/or augment the performance of one or more parent devices and is suitable for use or capable of functioning, whether or not it is packaged, labeled, or sterilized. In 2017, the FD&C was amended to classify a device accessory, when used as intended, based on its specific risks and require it to meet the level of regulatory controls respective to these risks. The amendment also created new pathways to enable a different classification for an accessory from its parent device. Accessories may be included in the parent device application or may be submitted separately (Food and Drug Administration 2017).

MULTIFUNCTIONALITY OF DEVICE PRODUCTS

Multifunctionality device products have at least one device function and one or more "other functions." The "other function" may or may not be regulated as a medical device. Multifunctionality device products that contain both a software non-device function and a device function may be assessed for the impact of the software non-device function on the safety and effectiveness of the device function. However, the software non-device function is not subject to review simply because it is part of a multiple function device product (Food and Drug Administration 2020). Software developers may design their multifunctionality device products to maintain a separation between regulated and non-regulated modules.

CONVENIENCE KITS

A convenience kit is "two or more different medical devices packaged together for the convenience of the user" (21 CFR 801.30(a)(11)). The convenience kit consists of two or more different medical devices that are packaged together and intended to remain packaged together and not to be replaced, substituted, repackaged, sterilized, unsealed, unwrapped, or otherwise processed or modified before being used by an end user. The end user may be an individual using the device on or on behalf of a patient, e.g., the patient, a caregiver, a healthcare practitioner, or a medical technologist. Convenience kits are themselves devices that have specific UDI requirements. Individual devices packaged within the immediate container of a convenience kit are excepted from the UDI labeling requirements, provided that a UDI is on the label of the immediate container of the convenience kit. For further information, refer to FDA guidance (Food and Drug Administration 2019).

DEVICE SUBMISSIONS

510(k) (Premarket Notification)

Class I and II devices not exempt from a 510(k) are required to notify the FDA of their intent to market a medical device for the first time at least 90 days in advance. This premarket notification or 510(k) submission is intended to demonstrate that the device to be marketed is safe and effective by establishing that it is substantially equivalent to a legally marketed (predicate) device (Section 513(i) of the FD&C Act (21 U.S.C. § 360c(i))). The FDA issues an order (510(k) clearance) within 90 days of receipt of the submission upon determining that the device meets its criteria for SE (Section 513(f)(1) of the FD&C Act (21 U.S.C. § 360c(f)(1))) market clearance

(Sections 510(k) and (n) of the FD&C Act (21 U.S.C. §§ 360(k) and (n))). A 510(k) submission is also required before making a significant change to a marketed device that may affect its safety or effectiveness. A significant change or modification could relate to the device design, material, chemical composition, energy source, manufacturing process, or intended use. The FDA maintains a monthly listing and also a searchable database of 510(k)s cleared by the FDA. Search results from the database can be a helpful tool for device classification and determination of a predicate.

PMA (Premarket Approval)

All Class III devices are subject to premarket approval requirements before they are placed on the market. Premarket approval requirements vary depending on whether the device was first distributed commercially on or after the May 28, 1976, amendments. Class III pre-amendments devices do not require a PMA to market until 30 months after the promulgation of a final classification regulation or until 90 days after the publication of a final regulation requiring the submission of a PMA, whichever period is later. The FDA may allow more than 90 days after the promulgation of a final rule for the submission of a PMA. Post-amendment devices that the FDA determines are substantially equivalent to pre-amendments Class III devices are subject to the same requirements as their predicate. Post-amendment devices determined by the FDA to be not substantially equivalent to a currently marketed Class I or II predicate device are "new" devices and fall automatically into Class III. These devices must have an approved premarket approval application or be reclassified into Class I (general controls) or Class II (standards) before such devices can be marketed. Clinical studies in support of an unapproved Class III device are subject to the investigational device exemption (IDE) regulations (21 CFR 812 for general devices or 21 CFR 813 for intraocular lenses).

Changes to a Class III device that affect safety and effectiveness require a PMA supplement. However, changes that involve modifications to manufacturing procedures or methods of manufacture require a 30-day notice or, where the FDA finds such notice inadequate, a 135-day PMA supplement. The FDA maintains a searchable database of marketed devices with premarket approval by the FDA. Search results from the database can be a helpful tool for device classification and determination of a predicate.

De Novo Classification Request

The De Novo process provides a pathway to novel medical devices for which there is no legally marketed predicate device. It is recommended to submit a pre-submission request prior to submitting the De Novo request to the FDA. The FDA makes a risk-based evaluation for the classification of the device into Class I or II, allowing their regulation through general controls alone, or general and special controls, while providing reasonable assurance of safety and effectiveness for the intended use.

Humanitarian Device Exemption (HDE)

The Humanitarian Device Exemption provides a regulatory pathway for Class III devices that are intended to benefit patients with rare diseases or conditions that affect fewer than 4,000 individuals in the United States per year. The HDE designation exempts a device from demonstrating that the device is effective for its intended purpose through a scientifically valid clinical investigation. The HDE application is expected to provide a risk-benefit analysis, information supporting device safety, information to show there are no comparable marketed devices or alternative forms of treatment, and the need for an exemption to bring the device to market. However, a human use device requires IRB approval for its use per the FDA-approved indication. The HDE is granted by the FDA's Office of Orphan Products Development (OOPD).

THE BREAKTHROUGH DEVICES PROGRAM

The breakthrough designation allows manufacturers developing Class III devices an opportunity to interact with the FDA's experts through several different program options to efficiently address topics as they arise during the premarket review phase. These options include sprint discussions; requests for discussion on a data development plan; and requests for a clinical protocol agreement that can help manufacturers receive feedback from the FDA, identify areas of agreement in a timely way, and prioritize review on future regulatory submissions, including Q-submissions, Investigational Device Exemption (IDE) applications, and marketing submissions.

Breakthrough designation requests for a device may be made through a Q-submission at any time prior to sending any of the aforementioned marketing submissions. Two criteria must be met to obtain breakthrough designation. The first criterion requires the device to provide for more effective treatment or diagnosis of life-threatening or irreversibly debilitating human diseases or conditions. The second criterion requires the device to either represent a breakthrough technology, have no existing marketed alternatives, offer significant advantages over existing marketed alternatives, or be in the best interest of patients.

DEVICE REGISTRATION AND LISTING

Annual registration with the FDA is required of foreign and domestic facilities or establishments that are involved in the production and distribution of medical devices intended to be placed on the US market. The FDA also requires registered establishments (except initial importers) to list all their devices and the activities that are performed on these devices. The purpose of the registration and listing requirements is to increase the nation's ability to prepare for and respond to public health emergencies.

The FDA manages a website that tabulates the requirements for registration and listing based on the type of activity performed at that establishment.

UDI AND GUDID

The unique device identification system was established by the FDA to identify certain medical devices sold in the United States so they may be traced from manufacturing through distribution to patient use.

Device manufacturers and labelers are required to include a unique device identifier on device labels and packages, except where the rule provides for an exception or alternative. The FDA supports manufacturers and labelers with exceptions and questions related to UDI and GUDID at the UDI Help Desk. A link to this help desk is available on the FDA website.

A unique device identifier (UDI) is a machine-readable code that is also required to be a human-readable unique numeric or alphanumeric code, and generally consists of the following parts:

1) **Device identifier (DI)**, which identifies the specific version or model and labeler of the device; and
2) **Production identifier (PI)**, which may contain one or more of the following device-specific pieces of information
   a. Lot or batch number
   b. Serial number
   c. Expiration date
   d. Date of manufacture
   e. Distinct identification code required by §1271.290(c) for a human cell, tissue, or cellular and tissue-based product regulated as a device

This UDI is also required to be loaded into the Global Unique Device Identification Database (GUDID).

Exceptions to the UDI requirement under §§ 801.2 are provided by §§ 801.30, 801.45, and 801.128(f)(2). A selection of these exceptions is summarized below:

Direct Marking

Re-usable devices requiring a UDI and intended to be reprocessed require either a machine-readable or human-readable UDI to also be directly marked on the device. The manufacturer can object to direct marking of their device after documenting it in the device design history file when it:

1) Interferes with the safety or effectiveness of the device
2) Is not technologically feasible
3) Is a single-use device subject to additional processing for an additional single use
4) Has been previously marked

Class I Devices

The UDI of a Class I device is not required to include PIs. Simply including a universal product code (UPC) on the label and packages is sufficient to meet the UDI requirements.

Some Class I devices exempt from the 21 CFR 820 do not require a UDI on the label or package.

Standalone Software

Standalone software distributed in packaged form must also include a UDI in both plain text and AIDC on the label and device package. If the software is downloaded, it may display a plain text statement of the UDI whenever the software is started or through a menu command.

Kits

A convenience kit requires a UDI on the kit's label. However, the individual devices inside the kit do not need a UDI. If a device in the kit is also marketed separately (as a replacement or otherwise), then it would require a UDI on the base package label.

Combination Products

A single entity combination product labeled with a National Drug Code (NDC) number does not require a UDI.

The device constituent part of a co-packaged combination product may have both an NDC code and UDI on the combination product label. However, the device constituent part itself is excepted from a UDI label only if it is packaged within the immediate container of the combination product.

Device Accessories

Medical device accessories are treated in the same way as devices, including the exceptions.

QUALITY MANAGEMENT SYSTEM (QMS)

Manufacturers of medical devices are responsible for following quality system requirements (QSRs). Certain low-risk devices are codified in 21 CFR 862 to 892 and exempt from these cGMP requirements. The exemptions do not apply to record keeping (21 CFR 820.180) and complaints (21 CFR 820.198). Medical devices manufactured under an IDE are not exempt from design control requirements under 21 CFR 820.30 of the QSR.

The FDA prescribed the current good manufacturing practices (cGMPs) for devices on December 18, 1978, in 21 CFR 820, as authorized by Congress under section 520(f) the FD&C Act. These cGMP practices included requirements such as the methods used in, and the facilities and controls used for, the manufacture, packing, labeling, storage, and installation of medical devices. Based on findings from two studies conducted in 1990, the law was revised again to modernize the medical device cGMPs under the Safe Medical Devices Act (SMDA) of 1990. These studies concluded that a significant proportion of device recalls could have been prevented by adequate implementation of design controls. The Agency further observed that the design-related defects were independant of the underlying device technology and complexity, and observed a diverse range of devices were affected, such as pacemakers, ventilators, chair lifts, administration sets, and software used to operate devices (see the preamble (pages 52602–52654) to the Quality System regulation (61 FR 52602)). The SMDA provided the FDA with the authority to add preproduction design controls to the cGMP regulation. After a significant public comment period, the initial proposal to the device cGMPs, published in the Federal Register on November 30, 1990 (55 FR 49644), underwent several revisions and was finally published in the 21 CFR 820 on June 1, 1997, as the Quality Systems Regulations (QSRs). During this review period, the Agency considered and adopted applicable public comments and international standards (draft ISO 13485). The avid and interested reader is directed to the preamble (pages 52602–52654) to the Quality System regulation (61 FR 52602) (Food and Drug Administration 1996) for valuable insights into the meaning and intent of the QSR.

The QSR is applicable to manufacturers that intend to commercially distribute their devices or functioning accessories that are suitable for use. The QSR does not prescribe quality requirements for any specific device or device type due to their wide range of risks, technological complexities, manufacturing processes, and diverse supply chains. The FDA expects device manufacturers to use the QSR as a framework to develop and continuously improve their device-specific manufacturing processes and procedures to meet current technological expectations. The flexibility of the QSR allows some latitude in implementing requirements, but this does not extricate the manufacturer from their obligations to market devices that are safe and effective. Manufacturers that use contractors or delegate work are still solely responsible for establishing and following processes and procedures to show that they design, produce, distribute, service, etc. their devices in compliance with the QSR.

The four pillars of the QSR are:

- Management
- Design and Development Controls
- Production and Process Controls
- Corrective and Preventive Actions

The first pillar, "Management," represents the expectation that executive management is responsible for establishing a quality policy, ensuring that it is understood and implemented, and maintaining an adequate organizational structure at all levels of the organization. This organizational structure should include an appointed management representative, who is adequately resourced with appropriate responsibility authority, and people to do the work. The appointed management representative is responsible for ensuring the establishment of the quality system and reporting on quality system performance to executive management. The suitability and effectiveness of the quality system should be reviewed by executive management at frequent and predefined intervals. The individual performing the audit should ensure compliance using established procedures and only audit work activities or processes for which they do not have direct responsibility. When required, quality measures should include a re-audit of deficiencies and corrective actions. Executive management should be provided with a written quality audit report for review. Sufficient personnel should be hired with the necessary education, background, training, and experience. Procedures should be established to document and identify training needs and ensure personnel are adequately trained. Training should make personnel aware of defects and errors that could be encountered as part of their job and also device defects that could occur from improper job performance.

Design and development controls are applicable to all Class II and III and the following Class I devices:

- Devices automated with computer software
- Catheter, tracheobronchial suction (868.6810)
- Glove, surgeon's (878.4460)
- Restraint, protective (880.6760)
- System, applicator, radionuclide, manual (892.5650)
- Source, radionuclide teletherapy (892.5740)

The intent of design controls is to build device design into the quality requirements for the manufacture of medical devices. The expectation is that once a device has been identified for development, a device-specific design and development plan (DDP) must be prepared to ensure that the design that will eventually be released to production adequately meets approved requirements. Adequate procedures should be established to ensure that this design translates to a device that properly performs its intended use and meets user needs. The DDP is not intended to be a retrospective documentation of device design. The DDP may be reviewed, updated, and approved as the design and development of the device progress. The DDP describes design and development activities, defines responsibility for implementation, and identifies and describes the interface between different groups or activities. Manufacturers are required to maintain procedures to design devices by defining and documenting design output in terms that allow an adequate evaluation of conformance to design input requirements. Design inputs are requirements that appropriately address the intended use of the device, and the needs of the user and patient. The design inputs should clarify incomplete, ambiguous, or conflicting requirements. The design output requirements are generally essential to the proper functioning of the device and must make reference to or contain acceptance criteria. Finished design output forms the basis of the device master record. The device master record (DMR) may include the device specifications for the production of an individual device, analogous to the recipe to bake a cake. The DMR may include instructions for device assembly, component specifications, manufacturing controls, sterilization procedures, device testing, and acceptance criteria of the finished device. Both design input and output requirements must be documented, reviewed, and approved by designated individual(s) before their release. The manufacturer is required to maintain procedures to verify device design. Design verification confirms by objective examination using measurable means and objective evidence that the design output meets the design input requirements. Design review is a documented, comprehensive, and systematic evaluation of the capability of the design to meet requirements. Design review may include an examination of the adequacy of the design requirements themselves and address problems. At appropriate stages of device development, manufacturers are required to formally document design reviews of the design results. Formal design reviews conducted at appropriately planned stages may be documented in the DDP. The participants at each design review must include representation from all functions involved at the design stage, at least one individual that does not have direct responsibility for the design stage being reviewed, and any specialists needed. The elements of design output and design verification are documented in the design history file (DHF). Manufacturers sometimes use a trace matrix to present and manage the relationship between design input and output requirements through design verification (e.g., design input and output are entered as the rows and columns, respectively, while specific design verification report results from the testing performed may be presented in the cells). Device manufacturers may use a risk management plan to aid in the proper design of their devices. The plan defines the responsibility and accountability of individuals with the required expertise to identify and manage possible hazards associated with the device design and production in both normal and fault conditions, including user errors. If any risk is judged unacceptable, it is reduced to acceptable levels. The preference is to reduce risk primarily by inherent device design using protective measures built into the device, its instructions for use and its manufacturing process, and may be followed up through labeling information for safety, such as warnings, and maintenance schedules, etc. Manufacturers are expected to monitor risk throughout the lifecycle of the device. Documentation from production, complaints, customer feedback, and market intelligence from similar devices is continuously evaluated to establish whether risks remain acceptable or if new hazards or risks are discovered. Design changes are required to be documented and approved upon appropriate risk analysis and verification before their implementation. Design validation ensures that the device conforms to its defined user needs and intended uses, and includes testing of production units under actual or simulated use conditions. Design validation is generally conducted through Human Factor Studies using approved methods and protocols on a device unit, lot, or batch manufactured under representative or actual commercial manufacturing processes. Design Validation reports are documented in the DHF. Changes in the medical device may require documentation of the design change in accordance with procedures that may result in a decision to repeat the design validation. Manufacturers must maintain procedures to ensure that the medical device design is correctly translated into production specifications to manage a design transfer for commercial production. The QSR applies to software used as a component of a medical device, software that is itself a medical device, and software used in the production of a medical device or in the implementation of the medical device manufacturer's quality system. Software validation and risk analysis are required where applicable.

Additional requirements under the CFR 820 are:

Production and process controls subsystem

- 820.50 Purchasing controls
- 820.60 Identification
- 820.65 Traceability
- 820.70 Production and process controls
- 820.72 Inspection, measuring, and test equipment
- 820.75 Process validation
- 820.80 Receiving, in-process, and finished device acceptance
- 820.86 Acceptance status
- 820.120 Device labeling
- 820.140 Handling
- 820.150 Storage
- 820.160 Distribution

MEDICAL DEVICE SAFETY

The FDA manages the safety of marketed medical devices through various methods. The Agency posts Medical Device Safety Communications for the public on its website as necessary to describe its current analysis of an issue and provide specific regulatory approaches and clinical recommendations for patient management. The Agency communicates safety concerns with medical devices used in health care facilities through letters to healthcare providers. Manufacturers and importers may be required to maintain records of all corrections and removals regardless of whether such corrections and removals are required to be reported to FDA. Manufacturers and Importers are required to report to the FDA any correction or removal of a medical device if the correction or removal was initiated to reduce a risk to health posed by the device or to remedy a violation of the provisions of section 519(g) of the Federal Food, Drug, and Cosmetic Act caused by the device that may also present a risk to health (21 CFR 806). A report must be made even if the event was caused by user error. Manufacturers and distributors may remove or correct a medical device that is in violation of laws administered by the FDA. A removal may also be required to protect the public from products that are deceptive, defective, or pose a risk of injury that cannot be corrected by a change in labeling. The FDA follows an enforcement policy that oversees device recall actions taken voluntarily by a company's correction or removal action (21 CFR 7). The Agency reviews the company-proposed recall strategy to address the problem. The Agency assesses the health hazard presented by the product and determines if the problem violates regulations and/or the law. The FDA may assign a recall classification to indicate its assessment of the relative degree of risk (I to III with decreasing degrees of risk). The Agency implements a recall strategy that can vary both in depth and breadth, depending on the specifics of the recall. The recall strategy is proportional to the assessed risk to adequately monitor and ensure the recall action has been effective. The Agency may determine that an establishment inspection is needed to identify the root cause of the problem and document additional observations and violations for possible regulatory action. While the recalling firm is responsible for determining whether its recall is progressing satisfactorily, the Agency may conduct effectiveness checks as part of its recall strategy.

The FDA communicates recall actions by posting information after classification and termination of the recall on its website in the Medical Device Recall Database (Medical Device Recalls 2002). In addition, the FDA may also post company press releases or other public notices about recalls, market withdrawals, and safety alerts that may potentially present significant risks to consumers or users of the product. The FDA has the legal authority to enforce the recall (Section 518(e) of the Federal Food, Drug, and Cosmetic Act; CFR 810) or ban (Section 516(a) of the Federal Food, Drug, and Cosmetic Act; 21 CFR 895.20) a medical device, although it has rarely needed to take these actions.

DIGITAL HEALTH

The scope of digital health is very broad and includes mobile health, health information technologies, wellness apps, wearable devices, telemedicine, clinical decision support software, personalized medicine, advanced robotics, machine learning, and artificial intelligence. Digital health uses a wide range of technologies such as data storage, data transmission, data manipulation, data portability, software, and artificial intelligence. Additionally, the disruptive nature of digital technology and its fast-evolving nature create significant challenges to the regulation of digital health. These technologies introduce a new and changing environment that requires carefully managed risk-based strategies to adequately balance regulation in an adaptive and innovative way, to not stifle their benefit through innovation. The CDRH specifically focuses its oversight on providing assurance of safety and effectiveness for medical device software functions based on risk to patient safety. The Agency uses this risk-based approach to determine whether it intends to regulate, exercise enforcement discretion, or not to regulate a specific software functionality. Examples of digital device functionalities that are not regulated by the FDA include those that are intented to provide access to electronic "copies" of other reference materials; educational tools for medical training of healthcare providers; access to commonly used reference information; automation of general office operations in a healthcare setting; and interactive health or wellness apps used by individuals. For an exhaustive list of functionalities along with their regulatory status, the reader is referred to the "Policy for Device Software Functions and Mobile Medical Applications" (Food and Drug Administration 2019). Additional examples of how the FDA intends to regulate different kinds of clinical decision support software functions are found in draft guidance titled "Clinical Decision Support Software" (Food and Drug Administration 2019). The very nature of digital health technology creates novel situations, requiring regulatory requirements to address interoperability, multifunctionality, and lack of physical location. Interoperability allows multiple devices to share information, raising cybersecurity and privacy risks, and requiring regulatory control to protect the devices from malware. Unlike physical devices, the regulatory requirements of software as a medical device apply to their function not their physical location . Hardware devices, for example may be connected through an accessory that collects information on their status or controls them. This accessory may also transmit information wirelessly to be received by digital software as a medical device on a mobile phone or even through the "cloud." The Agency does not regulate the mobile phone and clearly cannot regulate the "cloud" but regulates the software as a medical device through its functionality. Software as a medical device may be user-downloaded from a service that is also not regulated but must meet QSR requirements per 21 CFR 820.170 and 21 CFR 820.200, applicable to installation and servicing, respectively.

SUMMARY

Devices have been used throughout history to improve health outcomes with ever-increasing technological advancement, diversity, and complexity. As with any technology, there are risks and benefits that need to be balanced. The Agency attempts to maintain this balance through adequate requirements for regulatory compliance and control to ensure public safety while maintaining flexibility for technological advancement. The Agency classifies and regulates medical devices according to the risk during use. The responsibility of medical device manufacturers is to ensure that their devices are designed to be safe and perform to meet their intended use. As new technologies are introduced, the Agency focuses its oversight to address them. Manufacturers, as experts on their device and its technology, are best positioned to understand, assess, and ensure their devices are safe and function as intended. The expectation is that medical device manufacturers use principles of design controls to help them comply with regulatory requirements. Manufacturers that recognize this from an ethical perspective not only benefit from a favorable experience with the Agency but also from lower product liability and recall costs.

REFERENCES

94th U.S. Congress. (1976, May 28). Public Law. *S. 510*(90 Stat. 539), 94 - 295. Retrieved 02 07, 2021, from https://www.govinfo.gov/content/pkg/STATUTE-90/pdf/STATUTE-90-Pg539.pdf

Food and Drug Administration. (1996). Medical Devices; Current Good Manufacturing Practice (CGMP) Final Rule; Quality System Regulation. *Federal Register, 61*(195), 52601 - 52662. Retrieved 2 3, 2021, from https://www.fda.gov/science-research/clinical-trials-and-human-subject-protection/quality-system-regulations

Food and Drug Administration. (1998). Medical Devices; Exemptions From Premarket Notification; Class I. *Federal Register, 63*(21), 5387 - 5393. Retrieved 2 3, 2021, from https://www.govinfo.gov/content/pkg/FR-1998-02-02/html/98-2498.htm

Food and Drug Administration. (1998). Medical Devices; Exemptions From Premarket Notification; Class II. *Federal Register, 63*(13), 3142-3145. Retrieved 2 3, 21, from https://www.govinfo.gov/content/pkg/FR-1998-01-21/html/98-1485.htm

Food and Drug Administration. (2017). *Medical Device Accessories –Describing Accessories and Classification Pathways.* Retrieved 2 9, 2021, from https://www.fda.gov/media/90647/download

Food and Drug Administration. (2019). Retrieved 2 7, 2021, from https://www.fda.gov/media/109622/downloadFood and Drug Administration. (2019, 9 27). *Clinical Decision Support Software, Draft Guidance.* Retrieved 2 7, 2021, from https://www.fda.gov/media/109618/download

Food and Drug Administration. (2019). *Policy for Device Software Functions and Mobile Medical Applications.* Retrieved 2 7, 2021, from Policy for Device Software Functions and Mobile Medical Applications

Food and Drug Administration. (2019). *Unique Device Identification: Convenience Kits.* Retrieved 02 07, 2021, from https://www.fda.gov/regulatory-information/search-fda-guidance-documents/unique-device-identification-convenience-kits

Food and Drug Administration. (2020). *Multiple Function Device Products:Policy and Considerations.* Retrieved 2 9, 2021, from https://www.fda.gov/media/112671/download

Food and Drug Administration. (n.d.). *About the Digital Health Center of Excellence.* Retrieved 02 07, 2021, from https://www.fda.gov/medical-devices/digital-health-center-excellence/about-digital-health-center-excellence

Food and Drug Administration. (n.d.). *Digital Health Software Precertification (Pre-Cert) Program.* Retrieved 02 01, 2021, from https://www.fda.gov/medical-devices/digital-health-center-excellence/digital-health-software-precertification-pre-cert-program

Food and Drug Administration. (n.d.). *Food and Drug Administration.* Retrieved 2 7, 2021, from https://www.fda.gov/media/106331/download

Medical Device Recalls. (2002). Retrieved 2 3, 2021, from Food and Drug Administration: https://www.accessdata.fda.gov/scripts/cdrh/cfdocs/cfres/res.cfm#:~:text=Databases%20This%20database%20contains%20Medical%20Device%20Recalls%20classified,recall%20and%20again%20when%20the%20recall%20is%20terminated.

Medical Device Recalls. (2002). Retrieved 2 3, 2021, from Food and Drug Administration: https://www.accessdata.fda.gov/scripts/cdrh/cfdocs/cfres/res.cfm#:~:text=Databases%20This%20database%20contains%20Medical%20Device%20Recalls%20classified,recall%20and%20again%20when%20the%20recall%20is%20terminated.

Rados, C. (2006, January-February). FDA Consumer Magazine. *Medical Device and Radiological Health*, p. 8.

Sullivan, R. (1996, August). The Identity and Work of the Ancient Egyptian Surgeon. *Journal of the Royal Society of Medicine, 89*(8), 467-473. Retrieved 02 20, 2021, from https://www.ncbi.nlm.nih.gov/pmc/articles/PMC1295891/?page=2

Susruta & Bhishagrantna, K. L. (1907 - 1916). *An English translation of the Sushruta Samhita.* Calcutta, India: S. L. Bhaduri.

Suzanne, J. (2018). *The FDA History Vault.* (FDA voice) Retrieved 2 3, 2021, from https://youtu.be/ct7bqOVURg8

The Office of the National Coordinatot for Health Information Technology. (2014, April). FDASIA Health IT Report Proposed Strategy and Recommendations for a Risk-Based Framework. Washington DC: The Office of the National Coordinatot for Health Information Technology. Retrieved 2 7, 2021, from Food And Drug Administration (https://www.fda.gov/about-fda/cdrh-reports/fdasia-health-it-report): https://www.fda.gov/media/87886/download

6 FDA and Regulation of Biologics

Oliver V. Cácananta

INTRODUCTION

Biologics are at the forefront of scientific advancements in biotechnology, medicine, and healthcare. Prime examples are the rapid development and emergency-use authorization of messenger RNA vaccines to combat the COVID-19 pandemic and the proliferation of chimeric antigen receptor T cell therapy in cancer trials. The range of biological products is diverse, encompassing large, complex molecules. Vaccines, cells and tissues for regenerative medicine, transplantation, and gene therapy, blood and blood products, allergenics, and nucleic acids and most proteins are classified as biologics. Combination products can be considered biologics if the biological component has the primary mode of action; for instance, hematopoeitic stem cells with a delivery device would be considered a biologic. Ultimately, biologics are being developed to diagnose, prevent, and treat diseases, addressing many unmet medical needs.

The FDA's Center for Biologics Evaluation and Research (CBER) specifically has the authority to review and approve biologics for clinical use and commercial distribution. Initially, biologics were regulated by the Division of Biologics, created in 1937 by the National Institutes of Health (NIH). In 1972, jurisdiction was transferred to the FDA, and the division was renamed the Bureau of Biologics. The Bureau of Biologics then merged with the Bureau of Drugs in 1982 to form the Center for Drugs and Biologics. However, in a major 1987 reorganization, the FDA once again separated the oversight of drugs and biologics, establishing CBER as the authority for biologics, while the Center for Drug Evaluation and Research (CDER) assumed the responsibility for drugs. (Notably, the FDA has since transferred oversight of some therapeutic biologic products to CDER.)

The regulation of biologics played a key role in the history of the FDA. The first major federal law addressing medicinal products was the Biologics Control Act of 1902 (BCA), prompted by the deaths of over 20 children from tetanus contracted from contaminated diphtheria vaccines. The BCA authorized the regulation of the sale of vaccines, serums, toxins, and analogous products, establishing purity and safety standards, mandating the licensing of manufacturers and laboratories; and granting the federal government the authority to inspect these facilities. Subsequent

 DOI: 10.1201/9781315117911-6

key legislation includes the Pure Food and Drugs Act of 1906, the Public Health Service Act of 1944 (PHSA), and the Food, Drug, and Cosmetic Act of 1938 (FDCA).

The PHSA recodified the BCA, and Section 351 of the PHSA defined biologics, and established requirements for biologic manufacturers to apply for a product license before commercial distribution. Additionally, the statute specifically added the requirement of "potency" as a measure of clinical effectiveness. Section 361 of the PHSA granted the federal government for the first time to enforce quarantines to prevent the spread of communicable diseases. For example, this authority was invoked during the AIDS epidemic in the 1990s to regulate tissue transplantation and was cited in 2019 during the COVID-19 pandemic to implement federal quarantine guidelines. Section 361 also provides criteria for the regulation of human cells, tissues, and cellular and tissue-based products (HCT/P).

The FDCA, which replaced the Pure Food and Drugs Act, provides the FDA with regulatory authority over the safety of food, drugs, medical devices, and cosmetics, and serves as the foundation of biologic regulation. Major amendments to the FDCA include the Fair Packaging and Labeling Act (1966), the Medical Device Regulation Act (1976), and the FDA Modernization Act (1997). These laws continue to shape the framework for biologic product oversight.

DEFINITION AND PRODUCT CATEGORIES

Biologics and drugs are distinguished primarily by their molecular size and the processes used in their manufacture. Drugs are typically small molecules with low molecular weight and well-defined chemical structures, produced through chemical synthesis. Whether distilling cannabidiol from cannabis or formulating the active ingredients and excipients of a cough syrup in a state-of-the-art cGMP facility, the characterization and purification of drug products tend to be straightforward and highly controlled.

In contrast, biologics are macromolecules derived from biological systems, such as human, animal, plant, or microbial sources. These living systems introduce a high degree of complexity, making the development of chemistry, manufacturing, and control processes more challenging. Complete characterization of biologics is often difficult to achieve, whether scaling up production in cell culture systems or ensuring the absence of adventitious agents in tissue-based products. This inherent complexity poses unique obstacles to both the manufacturing and regulatory oversight of biologic products.

According to 21 CFR §600.3(h) and in Section 351(i) of the PHSA (42 U.S.C. 262(i)), the official definition of a biological product is

> a virus, therapeutic serum, toxin, antitoxin, vaccine, blood, blood component or derivative, allergenic product, protein, or analogous product, or arsphenamine or derivative of arsphenamine (or any other trivalent organic arsenic compound), applicable to the prevention, treatment, or cure of a disease or condition of human beings.

Viruses are generally defined as minute infectious agents that replicate in the living cells of organisms. They encompass a wide range of pathogens, from DNA viruses responsible for diseases like herpes and pox to RNA viruses that cause the common cold or rabies. However, from a regulatory perspective, the definition of what constitutes a virus extends beyond this traditional definition. According to the regulations, a product is analogous to a virus "if prepared from or with a virus or agent actually or potentially infectious, without regard to the degree of virulence or toxicogenicity of the specific strain used." Essentially, any product containing the minute living cause of an infectious disease, which may include bacteria, rickettsia, fungi, and protozoa, falls within this definition.

A toxin is a product

> containing a soluble substance poisonous to laboratory animals or to man in doses of 1 milliliter or less (or equivalent in weight) of the product, and having the property, following the injection of non-fatal doses into an animal, of causing to be produced therein another soluble substance which specifically neutralizes the poisonous substance, and which is demonstrable in the serum of the animal thus immunized.

An antitoxin is a product containing the soluble substance in the serum or other body fluid of an immunized animal which specifically neutralizes the toxin against which the animal is immune. Examples of toxin-antitoxins include the inactivated compounds produced by *Corynebacterium diphtheriae* and *Clostridium tetani,* which are used in diphtheria and tetanus vaccines, respectively. Diptheria antitoxins are derived from the serum of immunized horses while tetanus antitoxins can be derived from equine serum or from pooled human plasma. A product is analogous to a toxin or antitoxin if it "is intended, irrespective of its source of origin, to be applicable to the prevention, treatment, or cure of disease or injuries of man through a specific immune process."

Vaccines, which are biological preparations intended to provoke or boost an antigen-specific immune response for prophylactic or therapeutic immunization, have been regulated since the FDA's inception. They were the first biologics to be monitored and approved for use, with Congress enacting the BCA in 1902 in response to vaccine-related deaths. Vaccine development now encompasses a variety of strategies, from traditional methods using attenuated or inactivated microorganisms to more advanced approaches utilizing toxoids, vectors, biological subunits, or conjugates to stimulate immunogenicity. From the eradication of polio and smallpox to the mitigation of viral pandemics like influenza and coronavirus, vaccines have played and continue to play a critical role in safeguarding public health.

Whole blood, composed of red blood cells, plasma, and other constituents such as platelets, is a biologic, as are its various derivatives and transfusable components. Examples include albumin, immunoglobulins, clotting factors, fibrin sealants, proteinase inhibitors, recombinant and transgenic versions of plasma derivatives, blood substitutes, plasma volume expanders, radiolabeled or conjugated human or animal polyclonal antibody preparations, plasma-derived plasmin, and red cell reagents. Therapeutic serum, specifically identified in the legal definition of biologics, is blood without clotting factors and blood cells. Convalescent plasma with antibodies against the Ebola virus, for example, is therapeutic serum. A product is analogous to a therapeutic serum if it is "composed of whole blood or plasma or containing some organic constituent or product other than a hormone or an amino acid, derived from whole blood, plasma, or serum."

In 2020, the FDA finalized an amendment to the definition of biologics to add proteins to the list and clarify the framework under which they are regulated. A protein is

> any alpha amino acid polymer with a specific, defined sequence that is greater than 40 amino acids in size. When two or more amino acid chains in an amino acid polymer are associated with each other in a manner that occurs in nature, the size of the amino acid polymer will be based on the total number of amino acids in those chains and will not be limited to the number of amino acids in a contiguous sequence.

As previously noted, the FDA transferred regulatory authority for some therapeutic biological products to CDER, with the understanding that there would be regular interagency consultation whenever necessary. Many of these products are proteins. They include non-vaccine and non-allergenic immunomodulators; monoclonal antibodies for *in vivo* use; and proteins such as cytokines, growth factors, and enzymes derived from plants, animals, and microorganisms, as well as recombinant versions of these products, intended for therapeutic use, except for those that are assigned to CBER (e.g., vaccines and blood products).

CBER also has oversight for allergenic products, which are extracts used for the diagnosis and treatment of allergic diseases like hay fever and utilized in allergen patch tests and antigen skin tests, and the antimicrobial drug arsphenamine (also known as Salvarsan) and its derivatives, as well as any other trivalent organic arsenic compound. Arsphenamine is best known as a treatment for syphilis.

Lastly, a biosimilar product means that the product is decidedly comparable to a reference product, that is, a single, FDA-approved biologic. There might be minor variances in the inactive components; however, there are no clinically meaningful differences between them in terms of their safety, purity, and potency profiles.

With the rapid advancement in biotechnology and our growing understanding of biological systems, the range of biological products continues to expand. Categories now include cellular and gene therapies, tissue-based products, such as stem cells for transplantation, genetically engineered microorganisms, and even xenotransplantation. As the field evolves, so too must the regulations governing these products, ensuring safety, efficacy, and ethical oversight in this increasingly complex landscape.

FROM LABORATORY BENCH TO MARKET

Before any biological product can be made available to the public, it must receive FDA approval. The schematic below illustrates a high-level overview of the biological product lifecycle, highlighting key activities leading to commercialization.

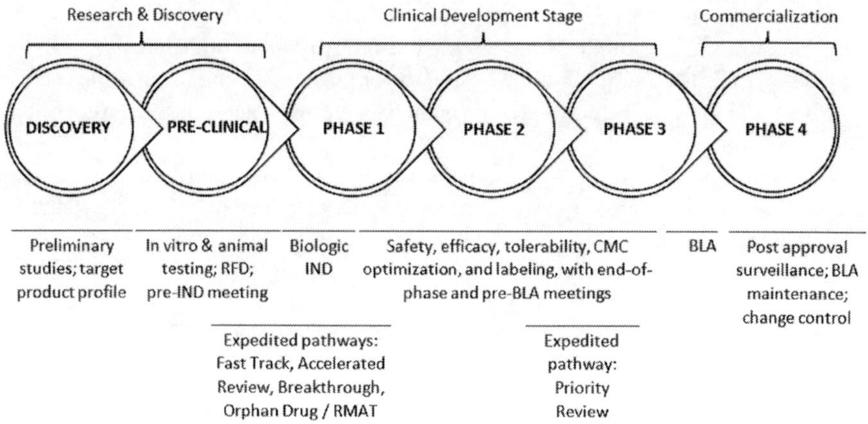

Research and Discovery

Every biological product begins with an idea, a question, or an observation. In the research and discovery phase, scientists apply the age-old scientific method to turn that inquiry into a practical application. They hypothesize how a product or technology might prevent, treat, or cure a disease, forming the basis for experimental testing to assess feasibility. Preliminary assays are developed with broad specifications, and animal testing is often conducted using small models like mice, rats, and guinea pigs, which are cost-effective at this stage. Meaningful results advance the project to the next stage for further testing. If results are unfavorable, the product returns to the drawing board or is discontinued altogether.

Pre-Clinical Development

After the discovery phase, and before a biologic can be tested in humans, sponsors must design a preclinical development program (PDP) that investigates the product's overall safety and effectiveness in non-animal and/or animal models. Given the promising results in earlier stages, animal testing may now involve larger, more relevant models such as pigs, dogs, and non-human primates. While PDPs for drugs typically focus on toxicity, pharmacodynamics, and pharmacokinetics, biologics present additional complexities due to the unpredictable nature of biological molecules and their interactions with living organisms. For biologics, PDPs may need to address concerns such as species cross-reactivity, immunogenicity, immunotoxicity, and carcinogenicity. Most preclinical studies must be performed in accordance with current Good Laboratory Practice (21 CFR §58), and the manufacturing process must begin to adhere to current Good Manufacturing Practice regulations (21 CFR §210 and §211).

The data generated from preclinical studies are crucial for the design of human clinical trials, as this is where safety is first evaluated. While the preclinical results are not always reliable predictors of human outcomes, they can help identify potential side effects and determine appropriate starting doses for Phase 1 clinical trials. Engaging the FDA at this stage is essential for obtaining feedback on the data and study design before filing an Investigational New Drug (IND) application.

One of the first considerations for sponsors is determining which FDA center is responsible for reviewing the biologic. As noted, there is some overlap regarding product oversight between CBER and CDER. If the biologic is part of a combination product, other FDA agencies, such as the Center for Devices and Radiological Health (CDRH) and the Office of Combination Products, might get involved. It is imperative, then, to establish communication with the proper premarket submission channel from the outset, as this affects the design of preclinical and clinical studies and the regulatory pathway for licensure.

If review and approval jurisdiction is unclear, sponsors have the option of first obtaining a preliminary assessment via the *Pre-Request for Designation* (pre-RFD) process. The pre-RFD provides informal, non-binding feedback regarding the regulatory classification of the product, as well as information about a product's assignment to the appropriate Agency Center. Typically, the pre-RFD suffices; however, if formal classification is necessary, for example, if there are disagreements regarding a combination product's primary mode of action, then the FDA's mechanism for that is the *Request for Designation* (RFD) process. The RFD decision is a binding resolution.

A sponsor may also engage the FDA in an INTERACT (Initial Targeted Engagement for Regulatory Advice on CBER Products) meeting to obtain preliminary informal, non-binding consultation for innovative investigational products at an early stage of development on issues that are not yet at the pre-IND meeting phase. According to the FDA,

> the INTERACT program generally consists of one informal and non-binding consultation and is intended for innovative investigational products that introduce unique challenges due to the unknown safety profiles resulting from the use of complex manufacturing technologies, development of innovative devices, or cutting-edge testing methodologies.

Once the correct Agency Center is identified, the next step is for the sponsor to request a meeting with the FDA to review and reach an agreement on the PDP studies needed to support human clinical testing. Conducting a formal pre-IND meeting significantly decreases the development timeline to product approval, as the sponsor can also obtain feedback on the format of the IND and the scope and design of Phase 1 clinical studies (21 CFR 312.82). Refer to the FDA draft guidance *Formal Meetings Between the FDA and Sponsors or Applicants of PDUFA Products* (2023) on the process of scheduling these meetings and the information required for meeting briefing packages.

Clinical Development: Phase 1

Following positive results from the preclinical stage, a sponsor advances to the clinical development phase, or first-in-human studies. Clinical development is typically divided into three sequential stages, each with increasing complexity and cost. Although the phases are often designed to follow one another, they can overlap in practice (e.g., a combined Phase 1/2 study evaluating safety and preliminary efficacy). Data from the discovery and preclinical phases, along with information on the nascent CMC processes, are compiled and filed with the FDA in an IND. An IND also includes a general investigational plan and an investigator's brochure (IB). However, an IB is not required in certain circumstances, such as an investigational-initiated study that has a right of reference to an existing manufacturer's IND application. Additionally, per 21 CFR 312.55 (a), an IB is not required for an IND where the investigator is also the sponsor, which is common in academic research.

The FDA assesses the IND to ensure the biological product is sufficiently safe to use in humans. If the FDA identifies concerns during the review, it may place the IND on "Clinical Hold" until the issues are resolved. Common concerns for biologics include insufficient characterization of master or working cell banks, potential toxicity of ancillary components, and inadequate CMC details. The review clock for an initial IND starts 30 days from the date the FDA receives the application. If no Clinical Hold is imposed within this timeframe, the sponsor may proceed with the clinical trial.

Once the IND is approved, the sponsor initiates the clinical trial, following approved protocols. This includes activating clinical sites and enrolling patients who provide informed consent. In addition to FDA oversight, the sponsor must secure approval from an Institutional Review Board (IRB), which ensures that clinical studies adhere to ethical research principles (21 CFR 56). Biological products are typically tested on adults first, for instance, before expanding to pediatric populations.

Phase 1 clinical studies emphasize safety and tolerability, generally enrolling 20 to 100 participants. Enrollment depends on the rarity of the disease or condition being treated or the complexity of the biologic. For instance, vaccines may require larger groups of healthy volunteers to assess adverse reactions across varying doses, whereas stem cell therapies in organ transplantation might enroll only a handful of subjects given the specificity of the procedure. Phase 1 studies may last for up to 12 months.

Between the preclinical stage and no later than Phase 2, sponsors should consider applying for FDA expedited pathways and incentive programs such as Fast Track, Accelerated Review, Breakthrough Designation, Orphan Drug Designation, and Regenerative Medicine Advanced Therapy Designation. These programs feature benefits such as rolling submissions, intensive guidance on development, financial incentives like tax breaks or marketing exclusivity, and approval based on surrogate or intermediate endpoints.

Before advancing to Phase 2, the sponsor must schedule an End-of-Phase 1 (EOP1) meeting with the FDA to discuss the trial results and obtain feedback on the next steps in clinical development.

Clinical Development: Phase 2

Phase 2 studies are designed to demonstrate efficacy or sufficient therapeutic activity, assuming Phase 1 has established a suitable dose range and identified no significant safety concerns. These studies typically enroll more subjects, up to 200 patients, and may last up to two years. Phase 2 trial designs vary, incorporating key elements such as therapeutic considerations, decision markers for advancing to Phase 3, and design characteristics (e.g., randomized versus non-randomized trials, or product versus placebo comparisons).

Often, Phase 2 trials are divided into two sub-phases: Phase 2a and Phase 2b. Phase 2a focuses on determining the optimal dose and dosing schedule, while Phase 2b emphasizes confirming efficacy in a larger patient cohort. The endpoints in Phase 2 can include measures of disease activity, symptom reduction, or biomarkers that correlate with clinical outcomes. Throughout the trial, safety continues to be closely monitored, with detailed tracking of adverse events to inform the design of future trials. This phase also presents an opportunity to identify specific patient subgroups that may benefit the most from the treatment, which can be crucial in refining the design for Phase 3 trials.

Phase 2 culminates in an End-of-Phase 2 (EOP2) meeting with the FDA, during which the trial's results, including efficacy, safety data, and plans for Phase 3, are reviewed. This meeting is essential for aligning the development strategy with regulatory expectations and ensuring the trial data supports the advancement of the drug into Phase 3.

Clinical Development: Phase 3

Phase 3 clinical development is the final and most extensive stage of clinical trials before regulatory approval, designed to confirm the efficacy and safety of a new drug or therapy on a large, diverse population. Typically, Phase 3 trials involve thousands of participants across multiple sites, often globally, to capture data from a broad range of demographics, comorbidities, and genetic backgrounds. These trials are usually randomized, controlled, and double-blinded to reduce bias and ensure robust data collection. The focus is on comparing the experimental treatment to either a placebo or the current standard of care. The trial duration can vary from months to several years, depending on the condition being treated and the required endpoints, which often include both primary outcomes (e.g., disease progression) and secondary outcomes (e.g., quality of life or biomarker changes).

Alongside clinical assessments, CMC play a critical role in Phase 3. Regulatory authorities require detailed CMC data to ensure the drug's consistency, purity, potency, and stability across different batches. The CMC section of the trial involves strict oversight of manufacturing processes, validation of production methods, and quality control tests to guarantee that the product administered to patients is both safe and effective. As production scales up for the larger patient population in Phase 3, sponsors must demonstrate that the drug can be manufactured reliably and that changes in the manufacturing process, such as shifts to commercial-scale production, do not affect the product's quality or efficacy. This is especially important for biologics, where even small variations in the manufacturing process can impact clinical outcomes.

Patient recruitment is a significant challenge due to the large sample size required. To meet these targets, trials are often multicenter or international, ensuring diversity and representativeness in the patient population. Throughout the trial, data on both efficacy and adverse events is rigorously collected, with ongoing monitoring by independent data safety monitoring boards to ensure patient safety. The identification and reporting of adverse events are key components in Phase 3, providing critical information for regulatory submissions. Regulatory approvals from health authorities such as the FDA or EMA, along with ethics reviews by Institutional Review Boards (IRBs), are required for the trial to proceed.

Statistical analysis in Phase 3 is designed to detect even small differences between the treatment and control groups, and trials are powered accordingly. Interim analyses may be conducted to assess early results, with the potential to stop the trial early if clear benefits or risks emerge. Final analyses focus on determining whether the drug meets predefined success criteria based on primary and secondary endpoints. If the trial is successful, the data collected forms the foundation for the Biologics License Application (BLA), with the CMC data ensuring that the drug can be manufactured consistently at a commercial scale. The data also informs drug labeling, including dosage, indications, contraindications, and potential side effects.

Phase 3 trials are costly and operationally complex, often requiring hundreds of millions of dollars to execute. They involve coordinating numerous trial sites, tracking patients, and maintaining consistency in data collection across diverse regions. Despite promising early-stage results, many drugs fail in Phase 3 due to unforeseen side effects or insufficient efficacy in a larger population. The integration of CMC data with clinical findings is essential for successful regulatory approval. When a drug succeeds in Phase 3, it paves the way for market release, making this phase a critical step in the overall clinical development and commercialization process.

A pre-BLA meeting is conducted with the FDA to discuss the application. Applying for Priority Review should be considered to reduce the review timeline.

Biologic License Application (BLA)

A BLA is a request to distribute a biologic product across state lines, regulated under 21 CFR 600. Any entity involved in the manufacturing of a biologic and responsible for ensuring compliance with these regulations can apply for a BLA. Before submitting the BLA, sponsors should arrange a formal pre-BLA meeting, participate in an advisory committee forum, and/or schedule a bioresearch monitoring inspection (BIMO). The timeline for these activities, along with their outcomes, influences the overall timeline for filing the BLA.

A formal pre-BLA meeting introduces the FDA reviewers to the sponsor's application and provides an opportunity to discuss the content and format of the marketing application. During this meeting, the sponsor should address the adequacy of preclinical and clinical studies, validation plans, appropriate statistical analysis methods, and any unresolved FDA concerns regarding the application. This meeting is crucial for ensuring that all data submitted are complete and in line with FDA expectations.

Applicants may also participate in an FDA advisory panel meeting, which provides independent expert advice on the scientific and clinical outcomes of the biological product. Advisory committees consist of external experts in science and medicine, as well as consumer, industry, and patient representatives. These meetings are open to the public, and while the panel's advice is non-binding, it serves as a strong indicator of potential product approval. The FDA ultimately makes the final decision, but advisory panel feedback often influences the outcome.

Scheduling a BIMO inspection is another critical step in the BLA process. BIMO inspections involve on-site inspections and data audits to ensure the integrity of the data submitted in the marketing application. This inspection is essential in confirming the quality of the clinical data, manufacturing processes, and compliance with Good Clinical Practice (GCP) standards. Successful inspections help ensure that the sponsor can move forward with filing their BLA.

Once these preparatory steps are completed, the sponsor can submit the BLA to CBER through the Document Control Center. Form 356h outlines the specific requirements for a BLA. Electronic submissions are now standard using the electronic Common Technical Document (eCTD) format. The BLA content includes the following:

- The administrative section including a table of contents and information on the applicant (Module 1)
- A summary of the nonclinical, clinical, and CMC information in the application (Module 2)
- A CMC section providing the quality management system, essential standard operating procedures, completed validations, specifications, analytical methods, stability, and labeling (Module 3)
- A nonclinical section including all relevant studies and results from the preclinical development program (Module 4)
- A clinical section that includes all safety and efficacy data as well as case report forms and serious adverse event narratives (Module 5)
- Supporting publications and literature.

The BLA then undergoes the FDA review process. Given the extensive interactions between the sponsor and the FDA during pre-BLA meetings, advisory discussions, and inspections, no surprises are expected at this stage. *SOPP 8401: Administrative Processing of Original Biologics License Applications (BLA) and New Drug Applications (NDA)* (Version 18 (2024), or its most current iteration) provides a detailed overview of the review and approval process.

POST-LICENSURE REQUIREMENTS

After a biologic product receives FDA approval, post-licensure requirements become essential to ensuring its continued safety, effectiveness, and quality throughout its lifecycle. These requirements encompass ongoing surveillance, post-marketing studies, quality control measures, and regulatory compliance activities designed to manage risks, confirm long-term benefits, and ensure product consistency as production scales up.

One of the most critical post-licensure requirements is ongoing safety surveillance. Even after extensive clinical trials, certain risks, especially rare adverse events or complications, may only become apparent when the product is used in a larger and more diverse population. The FDA employs several systems to monitor biologics after approvalBiologics Effectiveness and Safety (BEST) System, a complementary tool within the Sentinel initiative. These systems enable the FDA to monitor the long-term safety and effectiveness of biologics and to respond quickly to any emerging concerns.

The FDA may require sponsors to conduct Phase 4 post-marketing studies as a condition of approval, especially if there are outstanding questions about the product's risks, benefits, or long-term effects that could not be fully answered during pre-approval trials. These studies can focus on several key areas, includinglong-term safety and efficacy, assessment of rare adverse events by involving larger and more diverse populations, and risk mitigation strategies for products with known risks. The results of these post-marketing studies may lead to updates in product labeling, new warnings, or even withdrawal from the market if significant safety issues arise.

For biologics, which are often complex to manufacture, maintaining consistent product quality after approval is a significant challenge. The FDA implements a lot release program for certain biologics, which requires manufacturers to submit samples from each production lot, along with the associated test results, to the FDA for review. The FDA may conduct additional testing to ensure that each lot meets the required safety, purity, potency, and quality standards before it can be distributed. The lot release process is part of the larger Quality by Design (QbD) framework, which emphasizes the importance of designing manufacturing processes to ensure product quality from the start. Maintaining consistency across production batches is critical for biologics, and post-licensure lot testing and reporting requirements help ensure the product remains safe and effective throughout its lifecycle.

Manufacturers must continue to follow rigorous current Good Manufacturing Practices (cGMP) and validate any changes made to the production process post-approval. Changes might involve alterations to the manufacturing process, raw material sources, equipment, or facilities. Even minor process modifications can have a significant impact on biologics, given their complexity and sensitivity. The FDA requires manufacturers to submit detailed information about these changes for review to ensure that they do not affect the safety, potency, or purity of the biologic. These changes must be reported to the FDA as part of the BLA in one of three categories, depending on their impact on product safety or efficacy: major changes (Prior Approval Supplement), moderate changes (Changes Being Effected), and minor changes (Annual Report).

Post-licensure, the FDA continues to regulate product labeling and advertising to ensure that any claims made by the sponsor remain truthful, accurate, and consistent with the approved indications. Any new information gleaned from ongoing studies or adverse event reports must be reflected in updated labeling to inform healthcare providers and patients. Misleading advertising or promotion of off-label uses is strictly prohibited, and sponsors may face regulatory action if they do not comply.

If serious safety concerns arise after licensure, the FDA may work with the sponsor to recall the product voluntarily or, in more severe cases, mandate a market withdrawal. Sponsors are required to maintain a plan for managing product recalls, ensuring that any defective or dangerous biologics can be quickly identified and removed from the market to protect public health.

Post-approval inspections remain a critical part of ensuring continued compliance with FDA regulations. BIMO inspections can occur at manufacturing sites, clinical trial sites, or research institutions to verify that data submitted to the FDA are accurate and that the product continues to be manufactured under cGMP. Routine inspections help confirm that the sponsor adheres to the conditions of approval and that the biologic remains safe and effective for its intended use.

In summary, post-licensure requirements are vital to the lifecycle management of biologic products. They ensure ongoing safety surveillance, monitor product quality, assess the long-term efficacy of the biologic, and provide a mechanism for regulatory oversight. Through surveillance systems, post-marketing studies, lot release programs, and compliance inspections, the FDA works closely with manufacturers to protect public health and maintain the high standards set during the product's approval.

OTHER THINGS TO CONSIDER

- Emergency Use Authorization (EUA): If eligible, an applicant may file this to the FDA to facilitate the immediate availability and use of a product during public health emergencies. Such was the case with the COVID-19 pandemic.
- The FDA may consider allowing clinical studies to proceed without additional toxicology studies, depending on existing scientific and safety data available for the biologic, either from literature or related products. This should be discussed with the FDA during the Preclinical Stage.
- Approach clinical development with Adaptive Trial Designs in mind. That is another conversation with the FDA. The aim is to discuss the possibility of accelerating clinical trial decisions. The goal of these designs is to reduce the size and duration of the trial and demonstrate an effect if one exists. These adaptations are performed with close attention to statistical rigor.

RELEVANT REFERENCES

Two major sets of federal laws govern biologic regulations: (1) the United States Code (USC), which is the official compilation of general and permanent federal statutes passed by Congress; and (2) the Code of Federal Regulations, which codifies the USC issued by the executive departments and agencies of the federal government. Biologic researchers and manufacturers should refer to the following key regulations:

- Title 42 USC, Public Health and Welfare, Chapter 6A / Subchapter II / Part F:
  - Section 262, Regulation of Biological Products (coincides with Section 351 of the PHSA)
- Title 21 CFR, Food and Drugs:
  - Part 4: Regulation of Combination Products
  - Parts 50, 54, and 56: Protection of Human Subjects, Financial Disclosure by Clinical Investigators, and Institutional Review Boards, respectively
  - Part 58: Good Laboratory Practice for Nonclinical Laboratory Studies
  - Parts 210 and 211: Current Good Manufacturing Practice in Manufacturing, Processing, Packing, or Holding of Drugs; and for Finished Pharmaceuticals, respectively
  - Part 312: Investigational New Drug Application
  - Part 314: Applications for FDA Approval to Market a New Drug
  - Part 316: Orphan Drugs
  - Part 320: Bioavailability and Bioequivalence Requirements
  - Parts 600-680: Biologics
  - Parts 1270 and 1271: Human Tissue Intended for Transplantation, and Human Cells, Tissues, and Cellular and Tissue-Based Products, respectively

The FDA's website serves as a comprehensive resource, providing access to more than 2,600 guidance documents. These include drafts open for public comment and final documents, with over 600 specifically related to CBER. FDA guidance documents offer non-binding recommendations unless specific legal requirements are cited, serving as the agency's interpretation of policy on regulatory matters (21 CFR 10.115(b)). These documents cover various aspects of biologic regulation, from product design and labeling to manufacturing practices and compliance.

The strength of the USC, CFR, and FDA guidance lies in their comprehensive and prescriptive nature, while also leaving room for interpretation. This flexibility fosters collaboration and open communication with the Agency. As science and technology continue to advance, biologic regulations will likewise evolve, ensuring that innovation is balanced with safety and efficacy.

7 Combination Products

Evan B. Siegel

INTRODUCTION

Advances in drug, biologic, and medical device development deal with both single entities and combinations of each type of medical product. A desire for new modes of administration of therapeutics, enhancement of long-term delivery of drugs and biologics, new routes of administration, and commercial/competitive reasons have led to a veritable explosion of requests to the Food and Drug Administration (FDA or Agency) for review of experimental combination products. These developments continue to be enhanced by advances in technology and materials science, as well as the biotechnological sophistication of manufacturing processes. All of these factors have led to combination investigational drug, biologic, and device applications for human clinical testing. Eventually, these programs have led to New Drug Applications (NDAs) for marketing approval of small molecule drugs, Biologic License Applications (BLAs) for biologic products, and Premarket Notifications (510(k)) or Premarket Approval (PMA) applications for devices.[1]

WHAT IS A COMBINATION PRODUCT?

Under Title 21 of the Code of Federal Regulations (21 CFR) 3.2(e), a combination product is:

1. A product composed of two or more different types of medical products (i.e., a combination of a drug, device, and/or biological product with one another). The drugs, devices, and biological products included in combination products are referred to as "constituent parts" of the combination product;
2. A product comprised of two or more regulated components, i.e., drug/device, biologic/device, drug/biologic, or drug/device/biologic, that are physically, chemically, or otherwise combined or mixed and produced as a single entity;
3. Two or more separate products packaged together in a single package or as a unit and comprised of drug and device products, device and biological products, or biological and drug products);
4. A drug, device, or biological product packaged separately that according to its investigational plan or proposed labeling is intended for use only with an approved, individually specified drug, device, or biological product where

DOI: 10.1201/9781315117911-7

both are required to achieve the intended use, indication, or effect and where upon approval of the proposed product the labeling of the approved product would need to be changed (e.g., to reflect a change in intended use, dosage form, strength, route of administration, or significant change in dose); or

5. Any investigational drug, device, or biological product packaged separately that according to its proposed labeling is for use only with another individually specified investigational drug, device, or biological product where both are required to achieve the intended use, indication, or effect.

Examples of combination products where the components are physically, chemically, or otherwise combined include:

- A monoclonal antibody combined with a therapeutic drug
- A device coated or impregnated with a drug or biologic, e.g.,
    - Drug-eluting stent, pacing lead with steroid-coated tip, catheter with antimicrobial coating, condom with spermicide
    - Skin substitutes with cellular components, orthopedic implant with growth factors
- Prefilled syringes, insulin injector pens, metered-dose inhalers, and transdermal patches

Examples of combination products where the components are packaged together include:

- A drug or biological product packaged with a delivery device
- A surgical tray with surgical instruments, drapes, and lidocaine or alcohol swabs

Examples of combination products where the components are packaged separately but labeled for use together include:

- A photosensitizing drug and activating laser or other light source
- An iontophoretic drug delivery patch and controller

Examples of a wide variety of approved, licensed, or cleared combination products include:

- Neupro patch for the treatment of early Parkinson's disease
- Methylphenidate patch to treat attention deficit hyperactivity disorder in children
- Absorbable collagen sponge with genetically engineered human protein (device-biological product gel for surgical hemostasis)
- Iontophoretic transdermal system for fentanyl
- Paclitaxel-eluting coronary stent system
- Tositumomab and iodine I^{131} tositumomab

- Peginterferon alfa-2a in combination with ribavirin
- Iontophoretic drug delivery patch and controller
- Sirolimus-eluting coronary stent
- Photodynamic therapy involving a drug/device: methyl aminolevulinate cream with halogen light source

For the purposes of this chapter, and according to a broader regulatory developmental approach for multiple-entity medical products, the term "combination" is also used to connote drug/drug, biologic/biologic, etc. active principles, either incorporated into a single entity or co-packaged. This situation might be called "combined therapy" or "combinations of active ingredients."

THE US REGULATORY PROCESS FOR COMBINATION PRODUCTS

Section 3038 of the 21st Century Cures Act, enacted in December 2016 (P.L. 114–255) ("Cures Act"), substantially amended section 503(g) of the Federal Food, Drug, and Cosmetic Act (FD&C Act) (21 USC 353(g)), the principal section of the FD&C Act expressly addressing combination products.[2] General themes of these amendments include enhancing clarity, predictability, efficiency, and consistency of premarket regulatory expectations for combination products, including by ensuring that Agency components and staff coordinate appropriately on premarket review of these products, and that Agency thinking is aligned in conducting these reviews.

Regulatory review of combination products is complicated by the fact that while drugs are regulated primarily under section 505 of the Food, Drug and Cosmetic (FD&C) Act and reviewed by the Center for Drug Evaluation and Research (CDER), biologics are regulated under section 351 of the Public Health Service (PHS) Act[3] and reviewed by either CDER (therapeutic biologics, synthetic peptides and proteins [e.g., monoclonal antibodies], except for vaccines) or the Center for Biologics Evaluation and Research (CBER; vaccines, blood and blood products and medical devices used for collection, processing, administration of biological products and blood products or components, cell sorters, and in vitro tests and other medical devices involving retroviruses).[4, 5] The Center for Devices and Radiological Health (CDRH) regulates certain biological products and many combination products, including wound dressings with antibiotics or antiseptics, antimicrobial coated catheters and implants, bone cements with antibiotics, drug-eluting stents, photodynamic therapy, orthopedic implants with therapeutic drugs or biologicals, dermal replacement devices with living cells, and drug delivery devices (pumps, inhalers, pen injectors). Table 7.1 provides a delineation of product jurisdictions across centers. Combination product review at the FDA involves differing histories and philosophies at the various centers. For example, for evaluation of effectiveness devices require "valid scientific evidence," drugs require "substantial evidence," and biologics must be "safe, pure and potent." When a sponsor submits an NDA for a combination product for review, the cross-jurisdictional coverage may enter into the submission, review, and approval process. Recent (December 2020) FDA guidance on combination product reviews was expanded and clarified to include combination product agreement meetings

TABLE 7.1
FDA Centers with Jurisdiction over Drugs, Biological Products, and Medical Devices

Center for Drug Evaluation and Research (CDER)

- Non-biological (e.g., chemically synthesized) drugs
- Therapeutic biological products (except those specifically assigned to CBER) including:
  - Monoclonal antibodies for in vivo use
  - Chemically synthesized peptides and proteins intended for therapeutic use with 40 or fewer straight chain amino acids
  - Immunomodulators
  - Growth factors, cytokines, and monoclonal antibodies intended to mobilize, stimulate, decrease, or otherwise alter the production of hematopoietic cells in vivo

Center for Biologics Evaluation and Research (CBER)

- Biological and related products not regulated by CDER including:
  - Cellular products
  - Proteins intended for therapeutic use with more than 40 straight chain amino acids
  - Gene therapy products
  - Vaccines
  - Allergenic extracts
  - Antitoxins, antivenins, and venoms
  - Blood, blood components, plasma-derived products including recombinant and transgenic versions of plasma derivatives, blood substitutes, plasma volume expanders, human or animal polyclonal antibody preparations including radiolabeled or conjugated forms, certain fibrinolytics such as plasma-derived plasmin, and red cell reagents
- Medical devices related to licensed blood and cellular products including all HIV tests and test kits

Center for Devices and Radiological Health (CDRH)

- Medical devices including in vitro diagnostic products that are not regulated by CBER
- Radiation-emitting products

(CPAMSs), which provide a means of resolving issues and concerns between Agency reviewer and sponsors.[6] It is too early to assess the success of this additional avenue of communication and discussion in accelerating and rendering clearer the combination product review and approval process.

What Are the Roles of the Office of Combination Products?

By law, regulation, or FDA policy, the roles of the Office of Combination Products (OCP) are as follows:

- To serve as a focal point for combination product issues for Agency reviewers and industry.
- To develop guidance and regulations to clarify the regulation of combination products.
- To assign an FDA center to have primary jurisdiction for review of both combination and single entity (i.e., non-combination) products where the jurisdiction is unclear or in dispute.

- To ensure timely and effective premarket review of combination products by overseeing the timeliness of and coordinating reviews involving more than one Agency center.
- To ensure consistency and appropriateness of postmarket regulation of combination products.
- To resolve disputes regarding the timeliness of premarket review of combination products.
- To update agreements, guidance documents, or practices specific to the assignment of combination products.
- To submit annual reports to Congress on the Office's activities and impacts.

How are Combination Products Assigned for Review?

A combination product is assigned to an Agency center or alternative organizational component that will have primary jurisdiction for its premarket review and regulation. Under section 503(g)(1) of the FD&C Act, assignment to a center with primary jurisdiction, or a lead center, is based on a determination of the "primary mode of action" (PMOA) of the combination product. For example, if the primary mode of action of a device-biological combination product is attributable to the biological product, the Agency component responsible for premarket review of that biological product (CDER or CBER) would have primary jurisdiction for the combination product. If the PMOA of a device-biological product combination product is attributable to the biological product, for example, the center responsible for premarket review of such a biological product would have primary jurisdiction for the regulation of the combination product. The lead center for premarket review of the combination product also has the lead for ensuring compliance with CGMP regulatory requirements. Regardless of the PMOA, Agency components will coordinate as appropriate to enable efficient, effective CGMP regulatory oversight, including appropriate CGMP inspections.

The lead center is a manufacturer's primary point of contact. Manufacturers may also contact the OCP for assistance, as needed, in identifying appropriate contact points (including those in the lead center), resolving substantive issues, or otherwise facilitating interactions with the Agency and collaboration among Agency components, including centers and the Office of Regulatory Affairs (ORA).

As provided in section 503(g)(8)(C)(iv), as added by the Cures Act, the Agency will ensure that meetings between the FDA and sponsors are attended by review staff from each center as appropriate in light of the topics and purpose of the meeting.

What Investigational Application Should be Used for a Combination Product?

The FDA only needs one investigational application for a combination product. That application should include all information on the entire combination product. For example, if the investigation is for a drug-device combination product, the application should include the details on the drug and the device that typically would be submitted in an investigational new drug exemption (IND) and investigational device exemption (IDE), respectively. The FDA considers the entire combination product in determining the need for an investigational application. For example, in some investigational protocols, one part of the combination product is already approved but another part is new and raises safety risks. Therefore, the entire combination product raises a safety risk and would be under an investigational application. In some instances, both the drug and device are already approved for different indications, but when the two are used together, a new risk exists. In such cases, the combination product as a whole would be under an investigational application submitted to the lead center.

What Types of Marketing Applications are Required for a Combination Product?

Depending upon the type of combination product, its approval, clearance, or licensure may be obtained through submission of a single marketing application or through separate marketing applications for the individual constituent parts of the combination product. For small molecule drugs regulated under the FD&C Act, an NDA would be required; for biologic products regulated under the PHS Act, a BLA would be required; and for medical devices, a Premarket Notification (510(k)) or PMA application would be required. For most combination products, a single marketing application is sufficient for the product's approval, clearance, or licensure. In some cases, however, a sponsor may choose to submit two marketing applications for a combination product when one application would suffice. For example, a sponsor may choose to submit two applications to receive some benefit that accrues only from approval under a particular type of application (e.g., new drug product exclusivity, orphan status, or proprietary data protection when two firms are involved). In other cases, the FDA may determine that two marketing applications are necessary. For example, when one of the individual constituent parts of a combination product is already approved for another use and its labeling will need to be changed to reflect its new intended use in the combination product, the FDA may determine that two applications are necessary if the labeling of the already-approved product is subject to legal requirements different from those that will apply to the combination product. At the time this is being written, OCP is developing a guidance document for public review and comment addressing the factors that FDA expects to consider in determining whether single or multiple marketing applications should be submitted for a combination product.

How is a Product or Component of a Combination Product Classified?

The FDA has released a guidance that focuses on when a product may be classified as a drug or a device.[6] This guidance also addresses additional issues relating to product classification, including how to obtain a formal classification determination from FDA for a medical product and the status of prior Agency determinations concerning product classification.

If the classification of a product as a drug, device, biological product, or combination product is unclear or in dispute, a sponsor can file a request for designation (RFD) with OCP in accordance with 21 CFR Part 3 to obtain a formal classification determination for the product, as provided for under section 563 of the FD&C Act.

In reviewing an RFD, the Agency considers the information provided in the RFD as well as other information available to the Agency at that time. Generally, the Agency will respond in writing within 60 days of the sponsor's RFD filing, identifying the classification of the product as a drug, device, biological product, or combination product. If the Agency does not provide a written response within 60 days, the sponsor's recommendation respecting the classification of the product is considered to be the final determination.

The Agency may not modify a determination made under section 563 of the classification of a product or of the component of the FDA that will regulate the product except with the written consent of the sponsor or for public health reasons based on scientific evidence. However, the determination pertains only to the product described in the designation letter. A new determination may be appropriate if there is a change in, for example, an intended use or component of the product, or if the sponsor or Agency becomes aware of additional information that reveals that the mode of action differs from what was originally described in the RFD.

The FDA's determination of whether to classify a product as a drug or a device will be made based on the statutory definitions of these terms in the FD&C Act, as applied to the scientific data concerning the product that are available to the FDA at the time the classification determination is made.

The Agency interprets certain key provisions of the definition of device in the following manner:

- Products that meet the definition of device under section 201(h) of the FD&C Act also meet the definition of drug under section 201(g) of the FD&C Act, due to the broader scope of the drug definition.
- If a product is shown to meet both the drug and device definitions, the Agency tends to classify the product as a device.
- If a product meets the drug definition but there is uncertainty regarding whether it also meets the device definition, the Agency tends to classify the product as a drug.

Consequently, medical product classification determinations often focus substantially on whether the product meets the statutory definition of a device. The following discussion presents the Agency's current thinking on certain interpretive issues that arise with respect to the statutory definition of device.

A product that exhibits chemical action within or on the body of man may meet the device definition if the product "does not achieve its primary intended purposes through" such chemical action. Thus, if a product's chemical action contributes to an effect other than a primary intended purpose of the product, the product could fall within the scope of "device." In contrast, a product that depends, even in part, on chemical action within or on the body of man to achieve any one of its primary intended purposes would not be a device. In addition, if a product has multiple therapeutic effects, each of these would be a "primary intended purpose" of the product and the product would not meet the device definition if it achieves any one of these primary intended purposes through chemical action within or on the body of man.

Products that meet the drug definition, or both the drug and device definitions, may also meet the definition of biological product under section 351(i) of the Public Health Service (PHS) Act, which states that the term "biological product" means a virus, therapeutic serum, toxin, antitoxin, vaccine, blood, blood component or derivative, allergenic product, protein (except any chemically synthesized polypeptide), or analogous product, or arsphenamine or derivative of arsphenamine (or any other trivalent organic arsenic compound), applicable to the prevention, treatment, or cure of a disease or condition of human beings.

Some products that meet the drug definition, or both the drug and device definitions, and that also meet the definition of biological product, might be classified as biological products and be subject to licensure via a BLA under the PHS Act.

The Agency must classify products in accordance with the statutory definitions in the FD&C Act and the PHS Act. Sponsors often argue that their products should be classified in a certain way based on the classification of products they consider similar to their product. While the classification of similar products may help to inform the classification of the product at issue, the FDA believes that a case-by-case approach based on the specific characteristics of the product, including its intended use(s) and the current state of scientific knowledge at the time the classification determination is made, is necessary to ensure that products are classified properly under the applicable statutory criteria.[4]

Potential regulatory and legal issues the Agency is examining include determining when it may be appropriate to: (1) exercise enforcement discretion for products subject to an existing classification; (2) transfer products by determining that a currently approved application also meets the applicable requirements for another type of approved application (e.g., determining whether an approved NDA meets the requirements for an approved PMA); or (3) withdraw the existing approval for products and require new approvals for such products under the premarket approval authorities associated with the new classification. If the FDA determines that action should be taken to transfer or address the classification of any products, such efforts would be pursued in a transparent manner consistent with applicable legal requirements.

Under the definition of "device" in section 201(h) of the FD&C Act, a product (or a constituent part of a combination product) is not a device if it "achieve[s] its primary intended purposes through chemical action within or on the body of man or other animals." The FDA frequently receives questions from product developers concerning the Agency's interpretation of the term "chemical action" in this definition. An FDA guidance describes the Agency's interpretation of the term "chemical action."[7] However, the term "chemical action" must be read in the context of the statutory definition of "device" as a whole. The determination of whether a product meets the device definition does not depend solely on whether the product exhibits "chemical action." For example, a product that exhibits chemical action within or on the body of man may meet the device definition if the product "does not achieve its primary intended purposes through" that chemical action.[8]

Under the Agency's interpretation, a product exhibits "chemical action" for purposes of the device definition if, through either chemical reaction or intermolecular forces or both, the product:

- Mediates a bodily response at the cellular or molecular level, or
- Combines with or modifies an entity so as to alter that entity's interaction with the body of man or other animals.[9]

For example, when a product binds to a receptor through intermolecular forces, and thereby initiates or inhibits a signaling cascade in a certain type of cell within the body, that product exhibits "chemical action" because it mediates a bodily response at the cellular or molecular level through intermolecular forces. Similarly, a product that binds to a chemical agent through a chemical reaction, inhibiting the agent's effect on the body, exhibits "chemical action" because the product combines with an entity (the chemical agent) through chemical reaction so as to alter that entity's interaction with the body. For purposes of these factors, the term "chemical reaction" means the formation or breaking of covalent or ionic bonds, and "intermolecular forces" are electrostatic interactions or forces resulting from the interaction of localized, short-range electrical fields among atoms and/or molecules. Intermolecular forces include ion-dipole interactions, permanent dipole-based interactions, and induced dipole-induced dipole forces.

Under this approach, a product that exhibits covalent bonding, ionic bonding, or intermolecular forces does not exhibit "chemical action" under section 201(h) unless the product, through the covalent bonding, ionic bonding, or intermolecular interaction, mediates a bodily response at the cellular or molecular level or combines with or modifies an entity to alter the entity's interaction with the body. For example, dental devices that are used to fill hollowed out spaces in the teeth as a treatment of dental caries may bind to the tooth via covalent bonding or rely, in part, on intermolecular forces to change from liquid or paste to solid form. However, the product does not mediate a bodily response at the cellular or molecular level or combine with or modify an entity so as to alter that entity's interaction with the body through these chemical bonds. These products are therefore regulated as devices.

How Do I Write a Request for Designation (RFD)?

The mechanism for a sponsor to receive individual classification guidance from the OCP is via an RFD. The FDA has released a detailed guidance that outlines this process.[10] This guidance is intended to clarify the type of information the OCP recommends that a sponsor include in an RFD.

An RFD is also referred to as an applicant's letter of request (21 CFR 3.2(j)). It is a written submission to OCP. An RFD generally requests a determination of:

- The regulatory identity or classification of a product as a drug, device, biological product, or combination product, and/or
- Either the component of FDA that will regulate the product if it is a noncombination product, or which Agency center will have primary jurisdiction for premarket review and regulation if it is a combination product.

A letter of designation is FDA's formal response to an RFD and is a binding determination with respect to classification and/or center assignment that may be changed under conditions specified in section 563 of the FD&C Act and 21 CFR 3.9 in the regulations. It is not necessary to submit an RFD for every product. The FDA recommends submitting an RFD when the classification of a product or the Agency center to which it should be assigned is unclear or in dispute. Sponsors are encouraged to submit an RFD as soon as they have sufficient information for FDA to make a decision regarding classification or assignment of a product. The RFD should be submitted before filing any investigational or marketing application for the product. This will avoid a potential stay of the review clock if the classification or assignment of the product under review is determined to be unclear or in dispute during the review process.

If a sponsor has classification or assignment questions regarding multiple related products or product families that have different configurations, ingredients, and/or proposed uses or indications, the Agency recommends submission of a separate RFD for each product.

OCP may be contacted before submitting an RFD if the sponsor has any questions, including if there is uncertainty about the type of information to include or about the need to submit an RFD. It is not necessary to meet with OCP regarding most product classification and assignment questions or about an RFD submission. In most cases, the Agency will be able to make its determination based on a well-written RFD. Due to a mandatory short timeframe for reviewing RFDs, OCP generally will not grant a request for a meeting after the RFD is submitted. Meeting requests may be submitted by writing to OCP at their address or by e-mail to combination@fda.gov. The correspondence should be labeled as a "Meeting Request."

Within five business days of its receipt of an RFD, OCP will review the submission for completeness and determine whether the RFD contains the required information. OCP will then either send the sponsor an acknowledgement letter confirming the filing date of the RFD or notify the sponsor that the RFD was not filed, and identify the information needed to make the RFD complete for filing. For filed RFDs, the acknowledgement letter will also identify the date by which FDA plans to respond to the RFD. If FDA does not issue a designation letter within 60 calendar days of the filing of the RFD, the sponsor's recommendation for the classification or assignment of the product will become the designated classification or assignment.

A sponsor may withdraw an RFD by notifying OCP in writing any time after its submission and before the FDA issues its letter of designation. When the FDA issues its letter of designation, it constitutes an Agency determination that may only be changed in accordance with 21 CFR 3.9(b). Once the date-stamped letter of designation is issued by the FDA, a sponsor may not withdraw its RFD.

If a sponsor disagrees with OCP's jurisdictional determination, a sponsor may request reconsideration of a decision within 15 calendar days of receipt of the designation letter. A request for reconsideration cannot exceed five pages and cannot include any new information that was not contained in the original RFD. FDA will review and act in writing within 15 calendar days of its receipt of the request for reconsideration. If a sponsor wishes to present additional information or new data that were not presented in the RFD, a new RFD with this information may be submitted. OCP will consider this RFD a new submission.

According to 21 CFR 3.7(c), a sponsor is required to include the following information in an RFD, as applicable. RFDs are limited in length to 15 pages, including attachments.

1. The identity of the sponsor, including company name and address, establishment registration number, company contact person and telephone number.
2. A description of the product, including:
    i. Classification, name of the product and all component products, if applicable;
    ii. Common, generic, or usual name of the product and all component products;
    iii. Proprietary name of the product;
    iv. Identification of any component of the product that already has received premarket approval, is marketed as not being subject to premarket approval, or has received an investigational exemption, the identity of the sponsors, and the status of any discussions or agreements between the sponsors regarding the use of this product as a component of a new combination product;
    v. Chemical, physical, or biological composition;
    vi. Status and brief reports of the results of developmental work, including animal testing;
    vii. Description of the manufacturing processes, including the sources of all components;
    viii. Proposed use or indications;
    ix. Description of all known modes of action, the sponsor's identification of the single mode of action that provides the most important therapeutic action of the product, and the basis for that determination;
    x. Schedule and duration of use;
    xi. Dose and route of administration of drug or biologic;
    xii. Description of related products, including the regulatory status of those related products; and
    xiii. Any other relevant information.
3. The sponsor's recommendation as to which Agency component should have primary jurisdiction.

The original RFD and two copies should, as of this writing, be submitted with the envelope clearly marked as a "Request for Designation." The original RFD may be submitted by e-mail to combination@fda.gov in a common electronic format, such as PDF, in addition to two copies sent to the following address:

Office of Combination Products
Office of the Commissioner
Food and Drug Administration
WO32 Hub/Mail Room #5129
10903 New Hampshire Avenue
Silver Spring, MD 20993-0002

The OCP has established a new pathway for RFD advice,

Accordingly, FDA has enhanced the transparency and consistency of the process, now called the "Pre-Request for Designation (Pre-RFD) Program."[11] The "Pre-RFD" provides early guidance to sponsors of combination products.

The Pre-RFD process is available to provide informal, non-binding feedback regarding the regulatory identity or classification of a human medical product as a drug, device, biological product, or combination product. In addition, this informal process provides information about a non-combination or combination product's assignment to the appropriate Agency center for premarket review and regulation.

The above is a more flexible option than the RFD for a sponsor to receive a preliminary jurisdictional assessment, but which is not binding. Over time, informal methods of obtaining feedback have become increasingly customary with sponsors, and for some, even preferable to the formal RFD process. The guidance describes this structured process with clear recommendations for sponsors wishing to submit Pre-RFDs. It also provides the process for review of Pre-RFDs by FDA staff, the general timeframes for sponsors to receive feedback from OCP, and the process for scheduling teleconferences and meetings in relation to a Pre-RFD.

A Pre-RFD is a clear and concise written submission that a sponsor may make to OCP to request FDA's preliminary, nonbinding assessment of (1) the regulatory identity or classification of a product as a drug, device, biological product, or combination product, and/or (2) whether CBER, CDER, or CDRH will regulate the product if it is a non-combination product, or which of those Agency centers will have primary jurisdiction for the premarket review and regulation, if it is a combination product.

OCP will provide a written preliminary classification and/or jurisdictional assessment of the product based on the information provided in a specific Pre-RFD. The Agency's goal is to provide feedback within 60 calendar days of receipt of the basic information provided, and OCP will communicate with the sponsor during its review as needed.

Though Pre-RFD feedback is not binding on the Agency or sponsor, it is OCP's stated intent to provide the best advice possible based on the information provided in the Pre-RFD. OCP's review and written feedback will be made after an in-depth review, and it will contain a thorough and detailed rationale for its assessment. Moreover, the review will include involvement from the relevant Agency centers, as well as the Office of Chief Counsel when necessary.

Differences between the RFD and Pre-RFD process include length of submission and depth of regulatory analysis that a sponsor must provide. For example, RFDs cannot exceed 15 pages, but no such length requirement exists for Pre-RFD submissions. Moreover, for an RFD, a sponsor must provide an analysis of a product's classification, a primary mode of action (PMOA) analysis (if the RFD concerns combination product assignment), and a recommendation regarding Agency center assignment. Such information is optional in the Pre-RFD process. FDA will conduct the same type of PMOA analysis in a Pre-RFD as in a RFD regardless of whether a sponsor chooses to make a recommendation regarding the PMOA of its combination product in a Pre-RFD.

Acceptance and Filing Reviews for Premarket Approval Applications

For combination products containing devices that require PMAs, CDRH will make the determination if the information needed for safe and effective use of the device component is sufficient for approval, and therefore approval of the combination product as submitted.

An FDA guidance[12] provides detailed information and checklists for both acceptance of a PMA for filing review (if not accepted, a Refusal to Accept [RTA] notification will be sent to the sponsor); and the criteria for filing the PMA (if not filed, a Refusal to File [RTF] notification will be sent to the sponsor); and beginning the statutory review period. The PMA regulation (21 CFR 814.42 (e)) identifies the criteria, which, if not met, may serve as a basis for a refusal to file notification to the sponsor. Updates to the medical device laws and regulations, including MDUFA III, which went into effect in 2012, mandate FDA performance goals based on the timeliness of reviews of marketing applications. The PMA review team, in the case of a combination drug/device or biologic/device, will review the information provided to assure that if the device components are the subject of an RFD, it is, in fact, the same device for which information was submitted in the PMA. In addition, the team will check to see that the center to which the PMA was submitted (e.g., CDRH) is the same as that designated in the FDA response to the RFD.

User Fees

With regard to premarket authorization pathways, the FDA's current thinking is that a single application is generally appropriate for a combination product.[12] The marketing application type submitted should generally coincide with the PMOA of the combination product (e.g., a PMA, de novo, or 510(k) for a device-led combination product, an NDA or ANDA for a drug-led combination product, or a BLA for a biologic-led combination product).

One significant result of the assignment of regulatory jurisdiction for combination products is the user fee schedule that will be applied to the combination product for FDA review of marketing applications and for other FDA regulatory activities associated with the product. In general, combination products for which a marketing application is submitted will be subject to the fee associated with that type of application. Thus, combination products reviewed by CDRH will normally be subject to device user fees (under the Medical Device User Fee and Modernization Act, MDUFMA) and perhaps other fees (e.g., establishment registration fees) that may be assessed pursuant to legislative changes. Similarly, combination products assigned to CDER or CBER for review will be subject to Prescription Drug User Fee Act (PDUFA) fee requirements. PDUFA establishes NDA application fees, annual fees on establishments, and renewal fees on products. Sponsors may be eligible for fee waivers or reductions (e.g., for small businesses and orphan drug products) under PDUFA. Table 7.2 provides an overview of applicable fees and waivers for combination products.

As noted above, a sponsor may elect to submit two applications when only one would be required by FDA if it believes there is a benefit from doing so (e.g., new drug product exclusivity, orphan status, proprietary data protection when two firms are involved, additional market for a component combination product as a standalone product). The review of the additional application places an extra burden on FDA resources so the sponsor will generally be required to pay a fee for each application. Similarly, when the FDA requires two applications for a combination product, two application fees will be assessed to help defer FDA's review costs.

A PDUFA "barrier to innovation" waiver may be appropriate to reduce the additional fee burden associated with the FDA's requirement for two marketing applications. PDUFA provides for a fee waiver or reduction when the assessment of the fee would present a significant barrier to innovation because of limited resources available to the sponsor or other circumstances. The FDA has stated that "other circumstances" may exist in the relatively few instances where two marketing applications are required, and will consider the following as appropriate in determining eligibility for this fee reduction:

- The combination product (as defined in 21 CFR 3.2(e)) as a whole is innovative.
- FDA is requiring two fee-eligible marketing applications for the combination product.
- The applications only request approval of the two components of the combination product for use together. Applications that include independent uses of one or both components outside the combination product generally would not be eligible for this waiver. However, applications for combinations of already approved, independent products generally would be eligible if two applications are required for approval of the new combined use.
- The applicant does not qualify for a PDUFA small business waiver or have limited resources. Applicants who qualify for a PDUFA small business waiver receive a full waiver of their first PDUFA application fee. In addition, applicants with an innovative combination product who do not qualify for a PDUFA small business waiver but who have limited resources may be eligible for a standard PDUFA barrier to innovation waiver, which may provide a full waiver of the PDUFA application fee because of the applicant's financial need. More information about the standard barrier to innovation

TABLE 7.2
Combination Product Application Fees and Available Waivers or Reductions

Single Application/Fees[1]	Available Waiver/Reduction	Amount of Reduction
NDA/PDUFA fees; BLA/PDUFA fees	Waivers applicable to single PDUFA applications: • Small business[2] • Barrier to innovation • Necessary to protect the public health • Fees will exceed the anticipated present and future review costs incurred by FDA	PDUFA provides for a waiver of the fee for the first human drug application from a small business upon request. The FDA will either deny or grant a fee waiver or reduction request based on the PDUFA public health or barrier to innovation provisions after assessing the financial burden placed on the applicant by fees. If the FDA determines that the fees paid have exceeded the actual costs to review the application (using standard cost figures), the excess may be refunded upon request.
PMA, BLA, or 510(k)/MDUFMA Fees	Waivers applicable to single MDUFMA applications: • Small business • Humanitarian device exemption (HDE)[3] • BLA for a product licensed for further manufacturing use only • Third-party 510(k) • Any application for a device intended solely for pediatric use • Any application form a state or federal government entity	A small business may receive a one-time waiver of the MDUFMA fee that would otherwise apply to a PMA, BLA, PDP, or premarket report (premarket approval application for a reprocessed single use device – see http://www.fda.gov/cdrh/reprocessing/). There is no fee for an HDE application, a BLA for a product licensed for further manufacturing use only, an application for a device intended solely for pediatric use, or any application from a state or federal government entity.
Additional NDA or BLA[4]/PDUFA fees	PDUFA barrier to innovation	In applying the waiver to combination products requiring two PDUFA applications, the FDA would expect to reduce each PDUFA fee by half. In the case where two full PDUFA fees would otherwise be required, the total amount paid under this waiver would be equivalent to one PDUFA fee.

(Continued)

TABLE 7.2 (CONTINUED)
Combination Product Application Fees and Available Waivers or Reductions

Single Application/Fees[1]	Available Waiver/Reduction	Amount of Reduction
Additional PMA or 510(k)[4]/ MDUFMA fees	PDUFA barrier to innovation	In applying the waiver to products requiring a MDUFMA application and a PDUFA application, the FDA would expect to reduce the PDUFA fee by the amount of the MDUFMA fee. Thus, a sponsor would pay the full MDUFMA fee associated with the type of MDUFMA application, and a PDUFA fee reduced by the paid MDUFMA fee. The total amount paid would be equivalent to one PDUFA fee.

1. PDUFA applications (NDA or BLA) not requiring clinical data for approval are assessed half the fee that is assessed for applications that do require clinical data for approval. NDA or BLA supplements that require clinical data for approval are assessed half the full application fee; however, NDA or BLA supplements that do not require clinical data for approval are not assessed a fee.
2. The criteria for qualifying as a small business under PDUFA are different than those under MDUFMA. Under PDUFA, it is an entity that has fewer than 500 employees for the small business and its affiliates. Under MDUFMA, it is a firm with gross receipts or sales <$30 million. In FY 2007, firms with annual gross sales and revenues of $100 million or less, including gross sales and revenues of all affiliates, partners, and parent firms, may qualify for lower rates for PMAs, premarket reports, and supplements. A similar reduction may be available for subsequent fiscal years.
3. A Humanitarian Device Exemption (HDE) is an application to market a humanitarian use device (HUD). The HED application is similar to a PMA but exempt from the effectiveness requirements of the FD&C Act. An HUD is a "medical device intended to benefit patients in the treatment or diagnosis of a disease or condition that affects or is manifested in fewer than 4,000 individuals in the United States per year."
4. May be required by FDA or submitted at the discretion of the sponsor. FDA has indicated that when multiple applications are submitted at the discretion of the sponsor, FDA will take the sponsor's financial status into consideration in determining if a waiver or reduction will be granted.

> waiver is available on the FDA website. Applicants who believe they qualify are encouraged to explore their eligibility for the PDUFA small business or barrier to innovation waivers first.

FDA has indicated that it expects to consider several factors in determining whether a combination product is innovative for the purposes of this waiver, and notes that these are similar to the factors used in determining the eligibility of a product for expedited or priority review, where such factors are relevant to a determination of a combination product's innovativeness. A product need not be "life-saving" or for use in critical conditions to satisfy these factors, although the benefits of such innovation may be more evident in such cases. The factors include:

- The product addresses an unmet medical need in the treatment, diagnosis, or prevention of disease, as demonstrated by one of the following:
  - No approved alternative treatment or means of diagnosis exists, or
  - The combination product offers significant, meaningful advantages over existing approved alternative treatments.
- Factors such as whether one of the two applications includes a new molecular entity, has been designated as a priority drug or eligible for expedited device review, or has been granted Fast Track status,[14] may also be considered in determining whether a product is considered innovative for the purposes of this waiver.
- The existence of treatment alternatives would weigh against deciding that a product is innovative.

The sponsor of a combination product is urged to avail himself of the FDA consultation process in dealing with these and other issues.[13–16]

POST MARKETING SAFETY REPORTING FOR COMBINATION PRODUCTS

In April 2019, the FDA published a compliance policy guidance that addresses post-marketing safety reporting (PMSR) requirements for combination products that have received FDA marketing authorization, to ensure consistent and complete reporting while avoiding duplication.[17] In July of that year, the Agency released the final rule for PMSR.[18]

WHO IS SUBJECT TO THE COMBINATION PRODUCT PMSR FINAL RULE?

The following delineates the entities subject to the final rule:

- If one company is the applicant for a combination product that is marketed under a single application (e.g., a drug-eluting stent), this entity is the combination product applicant, and there are no constituent part applicants for the product.
- If one company holds the applications to market products for use together as constituent parts of a combination product (e.g., the company holds both the NDA for a specific photoactivated drug and the PMA for a device used to activate that drug where the two products compose a cross-labeled combination product), that company is the combination product applicant, and there are no constituent part applicants.

- If one entity holds the NDA for the drug constituent part and a separate entity holds the PMA for the device constituent part of that cross-labeled combination product, those two entities would be the constituent part applicants for that combination product, and there would be no combination product applicant for that product.
- A company is a constituent part applicant only if that entity holds an application to market that product as a constituent part of a combination product.
- A BLA or Device Application for a combination product that contains a drug constituent part. The combination product applicant must also comply with (21 CFR 4.102(c)(2)): Field alert reporting (FAR) requirements (see 21 CFR 314.81) and 15-day reporting requirements (see 21 CFR 314.80).
- An ANDA/NDA or Device Application for a combination product that contains a biological product constituent part. The combination product applicant must also comply with (21 CFR 4.102(c)(3)): Biological product deviation reporting (BPDR) requirements (see 21 CFR 600.14 and 606.171) and 15-day reporting requirements (see 21 CFR 600.80).

For combination product applicants, the safety reporting requirements under 4.102(b) based on the application type and under 4.102(c) based on the constituent parts of the combination product apply to the combination product as a whole. Accordingly, the entire combination product and each constituent part should be considered in determining whether a report is required and what information to include.

Various required reporting timelines are summarized below.

Timelines for Various Combination Product PMSR Requirements

Report Type	**Timeline For Reporting**
15-day reports	ANDA, NDA, and BLA combination products: no later than 15 calendar days from initial receipt of the information by the applicant (see 21 CFR 314.80(c) and 600.80(c)). Device Application combination products: no later than 30 calendar days from initial receipt of information by the applicant (see 21 CFR 4.102(c)).
Follow-ups to 15-day reports	ANDA, NDA, and BLA combination products: within 15 calendar days of receipt of new information (see 21 CFR 314.80(c) and 600.80(c)). Device Application combination products: no later than 30 calendar days from initial receipt of new information by the applicant (see 21 CFR 4.102(c)).
Five-day reports	No later than five workdays after the day that the applicant "becomes aware" that a reportable event(s) necessitates remedial action to prevent an unreasonable risk of substantial harm to the public health (see 21 CFR 803.53).

Timelines for Various Combination Product PMSR Requirements

Report Type	Timeline For Reporting
Death/serious injury/malfunction reports	No later than 30 calendar days after the day that the applicant receives information or otherwise becomes aware of the event (see 21 CFR 803.50).
Supplemental/follow-up reports to five-day/death/serious injury/malfunction report*	Within 30 calendar days of the day that the applicant receives information (see 21 CFR 803.56).
Field alert reports	Within three working days of receipt of the information by the applicant (see 21 CFR 314.81(b)(1)).
Biological product deviation reports	As soon as possible but not to exceed 45 calendar days from the date of acquiring information reasonably suggesting that a reportable event has occurred (see 21 CFR 600.14(c) and 606.171(c)).
Correction and removal reports	Within ten working days of initiating correction or removal (see 21 CFR 806.10(b)).

SPECIAL SITUATIONS FOR COMBINATION PRODUCTS

RMATs

Section 3034 of the 21st Century Cures Act (Cures Act) mandates that FDA issue guidance clarifying how FDA will evaluate devices used in the recovery, isolation, or delivery of regenerative advanced therapies (RMATs). The Agency issued a 2019 guidance that provides information about its current thinking regarding a wide range of concepts related to the regulation of devices used in the recovery, isolation, and delivery of regenerative medicine.[19]

The guidance addresses how the FDA intends to simplify and streamline its application of regulatory requirements for combination device and cell or tissue products; what, if any, intended uses or specific attributes would result in a device used with a regenerative therapy product to be classified as a Class III device; the factors to consider in determining whether a device may be labeled for use with a specific RMAT or class of RMATs; when a device may be limited to a specific intended use with only one particular type of cell; and application of the least burdensome approach to demonstrate how a device may be used with more than one cell type.

Consistent with section 503(g)(1)(B) of the FD&C Act, added by the Cures Act, and the Agency's approach to simplifying and streamlining the regulatory requirements for combination products in general, single entity combination products are generally reviewed under a single application. Devices intended for use with a specific RMAT may, together with the RMAT, be considered to comprise a biologic-led combination product and be evaluated for marketing under a BLA in the Center for Biologics Evaluation and Research (CBER), with consulting review by other center(s) as appropriate.

A combination product (biologic-device, biologic-drug, or biologic-device-drug) can be eligible for RMAT designation when the biological product constituent part is a regenerative medicine therapy and provides the greatest contribution to the overall intended therapeutic effects of the combination product (i.e., the primary mode of action of the combination product is conveyed by the biological product constituent part).

RMAT designation is granted for a specific therapeutic product and specific intended uses. RMATs represent a highly diverse group of products with distinct biological and physical characteristics. These characteristics, along with other factors such as target patient population (e.g., adult vs pediatric), intended use and conditions for use must be considered when determining which device, or devices, may be suitable for use with a specified RMAT or type of RMAT. These same factors will influence when a device may be limited to a specific intended use with only one particular cell-based RMAT or type of cell-based RMAT.

The Agency recognizes that a wide range of devices may be used in conjunction with an RMAT. For example, the devices can be simple, low risk devices, such as a manual surgical instrument (e.g., scalpel) for recovering cells and tissue. Devices used with RMATs may also be complex, higher risk devices, such as an automated cell collection system that selects and processes specific cells intended for immediate return back to the patient. Additionally, devices can be constituent parts of an RMAT that is a combination product.

During development, a given RMAT may be determined to have characteristics and use requirements that would allow it to be administered using a general class of devices (e.g., conventional syringe or catheter) or a specific subset of delivery device(s) with defined characteristics when such devices are used as labeled. Experience gained over time with a specific RMAT or class of RMATs in different use settings may also allow for identification of a general class or subset of delivery devices. If there is sufficient evidence that the safety and effectiveness of the RMAT would not be compromised when administered with the identified general class or specific subset of delivery devices, the RMAT could be approved with more general labeling rather than specifying a particular delivery device. In such cases, the labeling of the RMAT would specify the defined characteristics of the general class of devices (e.g., conventional syringe or catheter) or specific subset of delivery device(s) with which it is intended to be administered.

In Vitro Companion Diagnostic Devices

An in vitro companion diagnostic device provides information that is essential for the safe and effective use of a corresponding therapeutic product.[20] The use of an in vitro companion diagnostic device with a particular therapeutic product is stipulated in the instructions for use in the labeling of both the diagnostic device and the corresponding therapeutic product, as well as in the labeling of any generic equivalents of the therapeutic product. An in vitro companion diagnostic device could be essential for the safe and effective use of a corresponding therapeutic product to:

- Identify patients who are most likely to benefit from a particular therapeutic product;
- Identify patients likely to be at increased risk for serious adverse reactions as a result of treatment with a particular therapeutic product; or
- Monitor response to treatment for the purpose of adjusting treatment (e.g., schedule, dose, discontinuation) to achieve improved safety or effectiveness. FDA does not include in this definition clinical laboratory tests intended to provide information that is useful to the physician regarding the use of a therapeutic product but that are not a determining factor in the safe and effective use of the product.

In most circumstances, if use of an in vitro companion diagnostic device is essential for the safe and effective use of a therapeutic product, the in vitro companion diagnostic device and therapeutic product should be approved or cleared contemporaneously by FDA for the use indicated in the therapeutic product labeling.

The FDA encourages sponsors considering developing either the therapeutic or in vitro companion diagnostic devices discussed in the guidance to request a meeting with both relevant device and therapeutic product review divisions to ensure that product development plans will produce sufficient data to establish the safety and effectiveness of the in vitro companion diagnostic device/therapeutic product pair.

Over the last decade, the development of therapeutic products that depend on the use of a diagnostic test to meet their labeled safety and effectiveness claims has become more common. For example, such a test can identify appropriate subpopulations for treatment or identify populations who should not receive a particular treatment because of an increased risk of a serious side effect. One reason for increasing interest is the emergence of new technologies that can distinguish subsets of populations that respond differently to treatment. These technologies are making it increasingly possible to individualize, or personalize, medical therapy by identifying patients who are most likely to respond, or who are at lower or higher risk for a particular side effect.

When an appropriate scientific rationale supports such an approach, the FDA encourages the development of therapeutic products that depend on the use of approved or cleared in vitro companion diagnostic devices – several such in vitro companion diagnostic devices for use with corresponding therapeutic products have already been approved or cleared.

Inadequate performance of an in vitro companion diagnostic device could have severe therapeutic consequences. Such a device might fail analytically (e.g., by not accurately measuring the expression level of a protein of interest) or clinically (e.g., by not identifying those patients at increased risk for a serious adverse effect). Erroneous in vitro companion diagnostic device results could lead to withholding appropriate therapy or to administering inappropriate therapy. Therefore, the FDA believes that use of an in vitro companion diagnostic device with a therapeutic product raises important concerns about the safety and effectiveness of both the in vitro companion diagnostic device and the therapeutic product. The FDA will assess the safety and effectiveness of the in vitro companion diagnostic device as used with the therapeutic when a therapeutic product depends on the in vitro companion diagnostic device for its safe and effective use.

Applications for an in vitro companion diagnostic device and its corresponding therapeutic product will be reviewed and approved according to applicable regulatory requirements. The in vitro companion diagnostic device application will be reviewed and approved or cleared under the device authorities of the FD&C Act and relevant medical device regulations; the therapeutic product application will be reviewed and approved under section 505 of the FD&C Act (i.e., drug products) or section 351 of the PHS Act (i.e., biological products) and relevant drug and biological product regulations. FDA intends to review each in vitro companion diagnostic device submission within the context of, or in conjunction with, its corresponding therapeutic product, and FDA review of the test/therapeutic product pair will be carried out collaboratively among relevant FDA offices.

For a novel therapeutic product, an in vitro companion diagnostic device should be developed and approved or cleared contemporaneously to support the therapeutic product's safe and effective use (e.g., co-development). The results of the in vitro companion diagnostic device will be *essential* for the safe and effective use of the therapeutic product, and its use will be stipulated in the labeling of the therapeutic product (i.e., the therapeutic product is considered safe and effective *only* if used with the in vitro companion diagnostic device). Before approving the therapeutic product, the FDA will determine that the in vitro companion diagnostic device is properly validated and meets the applicable standard for safety and effectiveness or for substantial equivalence for the use indicated in the therapeutic product's labeling. Because the in vitro companion diagnostic device is typically considered essential to the safe and effective use of the therapeutic, the FDA does not believe it may approve a novel therapeutic product or new therapeutic product indication for use with an in vitro companion diagnostic device if the in vitro companion diagnostic device is not approved or cleared for that indication. Approval or clearance of the in vitro companion diagnostic device will ensure that the device has been adequately evaluated and has adequate performance characteristics in the intended population.

The FDA may decide that it is appropriate to approve a therapeutic product even though the in vitro companion diagnostic device for which it is labeled for use is not being approved or cleared contemporaneously. Two such scenarios are discussed below.

- New therapeutic products to treat serious or life-threatening conditions. FDA may decide to approve a therapeutic product even if its in vitro companion diagnostic device is not yet approved or cleared when the therapeutic product is intended to treat a serious or life-threatening condition for which no satisfactory alternative treatment exists and the benefits from the use of the therapeutic product with an unapproved or uncleared in vitro companion diagnostic device are so pronounced as to outweigh the risks from the lack of an approved or cleared in vitro companion diagnostic device.
- Already approved therapeutic products. FDA will generally not approve a supplement to an approved therapeutic product application to update the

> product's labeling to stipulate the use of an in vitro companion diagnostic device until the in vitro companion diagnostic device is approved or cleared. However, if the benefits from the use of the therapeutic product with an unapproved or uncleared in vitro companion diagnostic device are so pronounced as to outweigh the risks from the lack of an approved or cleared in vitro companion diagnostic device, the FDA will not likely delay approval of changes to the labeling of the therapeutic product until the in vitro companion diagnostic device is approved or cleared.

The FDA will apply a risk-based approach to determine the regulatory pathway for in vitro companion diagnostic devices, as it does with all medical devices. This means that the regulatory pathway will depend on the level of risk to patients, based on the intended use of the in vitro companion diagnostic device, and the controls necessary to provide a reasonable assurance of safety and effectiveness. Thus, the level of risk together with available controls to mitigate risk will establish whether an in vitro companion diagnostic device requires a PMA or a 510(k). FDA advises sponsors to consult early with the Agency on the likely regulatory pathway for an in vitro companion diagnostic device. Premarket review by FDA will determine whether the in vitro companion diagnostic device has adequate performance characteristics for its intended use.

If an in vitro diagnostic device is already legally marketed and the manufacturer intends to market its device for a new use as an in vitro companion diagnostic device for a novel therapeutic product, the FDA has stated that it would consider the new use of the in vitro diagnostic device with the novel therapeutic product a major change in the intended use of the device, raising new or additional questions of safety and effectiveness. Accordingly, an appropriate premarket submission (either PMA or 510(k)) for the new use must be approved or cleared for use with the novel therapeutic product.

New in vitro companion diagnostic devices intended to be used in the same manner as an existing approved or cleared in vitro companion diagnostic device (e.g., different manufacturer, different technological characteristics) will be reviewed under a PMA or a 510(k), as appropriate.

All diagnostic devices used to make treatment decisions in a clinical trial of a therapeutic product will be considered by the FDA as investigational devices, unless employed for an intended use for which the device is already approved or cleared. If used to make critical treatment decisions, such as patient selection, treatment assignment, or treatment arm, a diagnostic device generally will be considered a significant risk device under 21 CFR 812.3(m)(3) because it presents a potential for serious risk to the health, safety, or welfare of the subject, and the sponsor of the diagnostic device will be required to comply with the IDE regulations that address significant risk devices. In such cases, the FDA will expect the sponsor to conduct the trial under full IDE regulations.

If a diagnostic device and a therapeutic product are to be studied together to support their respective approvals (or clearance as appropriate for the diagnostic device), both products can be studied in the same investigational study if the study is conducted in a manner that meets both the requirements of the IDE regulations and the IND regulations (21 CFR Part 312). Information about the planned use of an in vitro companion diagnostic device and its use in clinical trials should be included in an investigational submission. This information will help FDA understand and provide advice on how the in vitro device will be used to enroll subjects into the trial and how the test will be validated for use. For therapeutic product INDs, the therapeutic product review center (CDER or CBER) will engage appropriate expertise from the diagnostic product review center (CDRH or CBER), and joint advice will be provided to the sponsor.

In addition, the FDA has stated that it will be helpful if both the in vitro companion diagnostic device product sponsor and the therapeutic product sponsor submit information about the proposed in vitro companion diagnostic device in a pre-IDE submission to the diagnostic review center. This will enable a more focused and in-depth discussion about the validation of the in vitro companion diagnostic device and will aid in planning for a complete and timely device PMA or 510(k). When appropriate, expertise from the relevant therapeutic review center will be included in the diagnostic review center meetings.

The FDA strongly encourages sponsors considering developing either the therapeutic product or the in vitro companion diagnostic device to request a meeting with both relevant device and therapeutic product review divisions as early in development as possible.

Contrast Imaging Indications

FDA has issued a guidance[21] to assist developers of medical imaging devices and drug/biological products that provide image contrast enhancement in determining:

- When the imaging device developer may add certain new imaging contrast indications to its device for use with already approved imaging drugs without a need for a modification of the drug labeling;
- When the imaging drug developer may add certain new imaging contrast indications to its drug for use with already approved imaging devices without a need for a modification of the device labeling; and
- What type of marketing submission(s) the imaging drug or imaging device developer should submit to FDA to request approval/clearance to add a new imaging contrast indication.

The purpose of this guidance is to describe the principles and provide examples under which drug or device developers can seek marketing approval/clearance to add new imaging contrast indications to the medical device using an already approved imaging drug.

Imaging device submission: When a new imaging device or device modification enables an approved imaging drug (i.e., at its approved formulation, dose, dosing regimen, rate, and route of administration) to be used for a new imaging contrast indication in a manner that is consistent with its approved indication, the FDA in most cases expects that a device submission from the device application holder alone would be sufficient to add the new imaging contrast indication to the device. The device submission would be either an original device application for a new device or supplemental device application for a modified device. Through this process, the imaging device developer would add the modified imaging contrast indication to the device labeling without the need to change the imaging drug labeling. For example, when new device software allows for the enhancement of the sensitivity and specificity of the same imaging site already identified in the approved imaging drug labeling, the drug is administered in accordance with its approved labeling, and the labeling does not need revision to ensure safety and effectiveness, the Agency believes that in most instances a device submission and device labeling change alone should suffice.

Imaging drug submission: In contrast, when an imaging drug modification (i.e., formulation, dosage, rate, or route of administration) enables the currently approved/cleared imaging device to be used for a new indication without a change to the device, the NDA/BLA holder should submit a supplement to request approval for such a change. For example, an NDA submission would be most appropriate for a drug reformulation that allows enhanced biodistribution to a new area but uses the same imaging software. FDA expects that, in most instances, submission of an NDA would be sufficient to add such an indication to the drug labeling without the need for a submission of a device application or a labeling change to the imaging device.

Imaging device and imaging drug submissions: In some instances, however, the device modification required to support the new imaging contrast indication may cause the drug and device to interact in a manner that affects the safety or effectiveness of the product(s); may change the imaging dosing, safety, or effectiveness information; or may produce labeling inconsistencies. In such circumstances, FDA expects that both the drug and device labels would need to be revised to ensure the safe and effective use of both products for the new imaging contrast indication.

Most imaging devices and drugs approved for use with a class of imaging products do not meet the definition of a combination product; e.g., when an imaging device is approved for use with a class of imaging drugs (e.g., gadolinium contrast) or when a drug is approved for use with a class of devices (e.g., MRI). If an imaging device is not a combination product, a manufacturer of an imaging device who intends to develop a new imaging contrast indication for use with a class of imaging drugs should submit a device application to CDRH for review. During the review process, CDRH will consult with CDER on issues including, but not limited to, the scientific/technical, risk/benefit, labeling, and potential interaction for the drug or drug class. Similarly, a manufacturer of an imaging drug who intends to develop a new imaging contrast indication for use with a class of devices should submit a drug application to CDER for review. CDER will consult with CDRH on all device-related issues.

In some instances, however, the use of a certain diagnostic imaging device and imaging drug may constitute a combination product. For example, a specific imaging drug used to bind receptors for imaging that uses a dedicated software algorithm may constitute a combination product. If a manufacturer has a combination product, the lead center determination will be in accordance with the primary mode of action regulatory provisions. For the developer of an imaging device wishing to add a new imaging contrast indication for a class of imaging drugs, the supportive clinical trial(s) should be conducted under an IDE with a submission to CDRH. Similarly, for the developer of an imaging drug wishing to add new imaging contrast indications for a class of devices, the supportive clinical trials should proceed under an IND with a submission to CDER. For a combination product, the submission should proceed under the lead center as determined by the product-specific primary mode of action, and typically the investigational application would be of the type administered by the lead center.

Glass Syringes for Delivering Drug and Biological Products

FDA has recognized International Organization for Standardization (ISO) Standard 11040-4 for glass syringes that are intended to deliver drugs or biological products. FDA has become aware of adverse events and product quality events related to connectivity problems when certain glass syringes are used with connecting devices, including connecting devices that conform to the FDA-recognized ISO 594-2 standard. "Connecting devices" include needles, needleless Luer connectors, adapters, and transfer units. Accordingly, FDA has determined that, for glass syringes, demonstrating conformity with the ISO 11040-4 standard alone does not ensure that the glass syringe can be properly connected to connecting devices. A recent FDA guidance identifies additional technical information that should be submitted by sponsors who seek to rely on conformity to ISO 11040-4 for a glass syringe product.[22] The recommendations provided in this guidance document are applicable to the sponsor of an IDE, humanitarian device exemption (HDE), 510(k), or PMA for the following glass syringe products, and to the sponsor of an IND, BLA, NDA, or ANDA for a drug or biological product that is delivered with such a glass syringe product, including a product that qualifies as a combination product:

- Needleless glass syringes prefilled with a drug or a biological product;
- Empty glass syringes co-packaged with a drug or a biological product and/ or connecting device; and
- Empty glass syringes intended for use with marketed drugs or biological products.

Pen, Jet, and Related Injectors for Use with Drug and Biological Products

The FDA has provided a draft guidance document to assist industry in developing technical and scientific information to support a marketing application for a pen, jet, or related injector device intended for use with drugs and biological products, including combination products.[23] The marketing application would typically be a 510(k) or a PMA application for the injector alone. For a combination product that includes the injector, the marketing application would typically be an NDA or BLA.

For purposes of this guidance, the term injector includes, but is not limited to, jet injectors, pen injectors, piston syringes, needle-free injectors, mechanically operated injectors, and injectors with computerized or electronic elements.

Pen, jet, and related injectors may provide an innovative approach to deliver drugs or biological products and may enhance safety, improve dosing accuracy, and increase patient compliance, particularly in self-administration settings. There are three basic groups of injectors: those intended for general use with a wide range of currently marketed drugs/biological products, those intended for use with a certain class of drugs/biological products or with a specific product line, and those intended for use with a specific drug/biological product. Depending upon the technology, an injector may be pre-filled with a drug/biological product for single or repeat dosing. For example, a given injector may be for single use only in acute intervention for treatment or prevention while another injector may be for repeat dosing as part of a long-term treatment regimen. Other injectors may be co-packaged with the drug or biological product for assembly by the patient or health care provider.

Injectors may be subject to different regulatory provisions depending on their intended use, technological characteristics, proposed labeling, and packaging. For example, most injectors are intended for general use with a wide range of drugs/biological products and are marketed alone. Typically, general use injectors have marketing clearance as devices under a pre-market notification (510(k)). Other injectors are intended for use with a specific drug/biological product and may be pre-filled with that product, co-packaged with that product, or separately provided with mutually conforming labeling. These injectors are typically combination products under 21 CFR 3.2(e). For most combination products, one marketing application is sufficient for the injector and the drug/biological product; this application is usually an NDA/BLA. When appropriate, however, FDA may determine that two applications are necessary (i.e., an NDA/BLA along with a device submission that is usually a 510(k)).

The draft guidance focuses on the scientific and technical considerations that may be relevant when developing a pen, jet, or related injector regardless of the regulatory provisions that apply (i.e., device, drug, or biological product provisions). It provides considerations applicable to a general use injector and to an injector intended for a product class, product line, or a specifically named drug/biological product. It also provides general content and format information for injectors to be reviewed under 510(k) submissions and for review as part of an NDA/BLA submission. It focuses specifically on scientific and technical information for the injector, which would be part of any marketing application. If a sponsor is proposing a new route, dose, or rate of administration for the drug or biological product, FDA expects that a CDER–CDRH collaboration will be necessary, as appropriate for the type of submission. The Agency recommends that the sponsor contact the CDRH review branch or the OCP for additional information.

Package Labeling for Injectable Medical Products

An FDA guidance provides industry with the Agency's recommendations on the selection of appropriate package type terms and selection of appropriate discard statements for injectable medical products for human use, packaged in multiple-dose, single-dose, and single-patient-use containers.[24] Specifically, this guidance provides FDA's revised definitions for single-dose and multiple-dose containers as well as for the new package-type term "single-patient-use container." These containers may be part of a drug, a biological product, or a combination product assigned to CDER, CBER, or certain combination products assigned to CDRH. Marketing applications for such products include: New Drug Applications (NDAs), Abbreviated New Drug Applications (ANDAs), Biologics License Applications (BLAs), Premarket Approval Applications (PMAs), Premarket Notifications under section 510(k) of the Federal Food, Drug, and Cosmetic Act (FD&C Act), and requests for classification submitted under section 513(f)(2) of the FD&C Act (De Novo request). Sponsors should determine the proper package type term ("single-dose," "multiple-dose," or "single-patient-use") for injectable medical products for human use, and use only the correct term for the package type throughout the labeling. FDA recommends that the appropriate package type term appear on all components of the labeling of injectable medical products for human use so the user will be able to easily identify the package type. This includes the container label and carton labeling, prescribing information, and where applicable, labeling intended for the patient.

NONCLINICAL RECOMMENDATIONS FOR SUCCESSFUL CHARACTERIZATION AND DEVELOPMENT OF COMBINATION DRUG PRODUCTS

Many pharmaceutical and biotechnology companies have successfully developed drug candidates (both small organic molecules or novel chemical entities and macromolecules or biologicals, both of which will be designated in this chapter as novel molecular entities [NMEs]) that have been approved for marketing by FDA or other regulatory authorities. Most of these NMEs have been studied as single therapeutic agents for the treatment of a human disease or disorder. The regulatory recommendations for the characterization and development of these NMEs are described in various FDA regulations, FDA guidance documents, and International Conference on Harmonisation (ICH) guidelines and have been documented elsewhere.[25–27] Table 7.3 summarizes these requirements.

For the successful treatment of many human diseases and disorders, however, a single therapeutic agent is not the "optimal" choice, and combination therapy is necessary to provide the desired biological or therapeutic response with an acceptable human safety or toxicity profile. The development requirements for a combination therapy or combination drug product (CDP) differ from those for a single agent since the pharmacologic activities and toxicologic profiles of the marketed drugs and/or NMEs in a CDP may be substantially different from those observed for each of the single agents alone. The changes in biological potency for a CDP may include, but are not limited to, synergism (desirable) or enhanced or prolonged pharmacologic activity in combination compared to the activity (extent and/or duration) of the single agents alone; additive effects (desirable) when the therapeutic effects of the single agents combined provide more effective treatment; or antagonism (usually undesirable) when the therapeutic effects of one or more of the single agents in a CDP are reduced when administered concurrently with other marketed drugs and/or NMEs.

TABLE 7.3
Recommended Pharmacology and Safety Pharmacology Activities for Combination Drug Products Containing Marketed Drugs and/or New Molecular Entities

Nonclinical Area	Recommended Activity
Pharmacology	
Marketed drug	Evaluate available pharmacology database and prepare appropriate summaries of results for inclusion in regulatory submissions.
NME	Conduct necessary pharmacology studies to define dose level and regimen for desired activity in animal models.
CDP with only marketed drugs	If considered desirable, conduct animal studies to evaluate pharmacology profile when components in a CPD are dosed using established clinical routes and frequency of administration for each drug.
CPD with marketed drugs and one NME	Conduct animal studies to evaluate pharmacology profile of CPD and to determine optimal dose levels and frequency of dosing for NME when administered concurrently with marketed drugs.
CPD with more than one NME	Conduct animal studies to evaluate pharmacology profile of CPD and to determine optimal dosage regimen for each NME when administered concurrently with another NME.
Safety pharmacology (SP)	
Marketed drug	Evaluate available safety pharmacology database and if necessary, conduct SP studies to meet current regulatory standards.
NME	Conduct recommended ICH SP studies (standard battery and where appropriate, supplemental battery).
CDP with only marketed drugs	If one or more marketed drugs has incomplete SP assessments or if the marketed drugs produce adverse SP effects for the same organ system, design and conduct appropriate SP studies on CDP to determine if SP results are adversely affected when marketed drugs are administered in combination.
CPD with marketed drugs and one NME	If SP studies on the individual components indicate possible adverse SP effects for the same organ system, design and conduct appropriate SP studies on CDP to further evaluate and extend the findings for that organ system.
CPD with more than one NME	If SP studies on the individual NMEs show adverse effects on the same organ system, design and conduct appropriate SP studies on the CDP to further evaluate and extend the findings for that organ system.

The changes in the toxicologic profile for a CDP may include, but are not limited to, reduction (desirable) of adverse effects because lower doses of one or more of the single agents in a CDP can be administered with a similar or improved pharmacologic profile; modification (desirable) of adverse effects caused by one or more of the single agents in a CDP by another marketed drug and/or NME; similar (acceptable) adverse effect profile for each marketed drug and/or NME in the CDP; or increased (undesirable) adverse effects because the safety or toxicity profile of one or more of the single agents in a CDP is changed when administered in combination with other agents.

In addition to defining and characterizing the pharmacologic and toxicologic profiles of the marketed drugs and/or NMEs in a CDP, another important area that needs to be evaluated is potential changes in the pharmacokinetic and drug metabolism profiles of each of the single agents to be included in a proposed CDP. Possible desirable or undesirable (depending on why the CDP is being developed) changes in pharmacokinetic profiles include, but are not limited to, increases or decreases for one or more of the single agents in a CDP in the rate and/or extent of absorption, in plasma protein binding characteristics, in the tissue distribution profiles, and in the rate and/or route of elimination. Changes in the extent and type of metabolism for any of the single agents in a CDP also may be desirable or undesirable. Increased metabolism may reduce the adverse effect profile of one marketed drug or NME but could also cause more rapid clearance and thus decrease the pharmacologic effectiveness of a second agent. Decreased metabolism may provide longer exposure to a pharmacologically active compound and thus produce enhanced effectiveness but could also cause accumulation of one or more of the single agents in a CDP in one or more tissues and/or organs and thus potentially cause increased adverse effects. Table 7.4 summarizes recommended pharmacokinetic, toxicokinetic, and drug metabolism suggestions for combination products containing marketed drugs and/or NMEs. Table 7.5 provides an overview of recommended nonclinical safety or toxicology support for such products.

NDAs for combination products must take into account these issues, and strategies for developing these products should include the content of the eventual marketing application stemming from the development program.

CLINICAL PHARMACOLOGY AND CLINICAL DEVELOPMENT OF COMBINATION PRODUCTS

Safety testing of the components of combination products should be integrated with planning for the clinical pharmacology studies and the clinical studies that will generate data supporting substantial evidence of safety and efficacy. The interaction of combination products with the human body is often more complicated than that of a single-entity drug, biologic, or medical device. Sponsors should strategically approach pharmacology and clinical testing as a continuum with a staged, integrated, and logical plan that will minimize expenditure of scarce funding, time, and resources, and yet satisfy the needs of the regulatory authorities to protect the public from the unknown effects of such products. Bioavailability, bioequivalence, and efficacy combined with acceptable safety in the human population of interest should be covered, with the NDA or BLA content foremost in mind.

An increased understanding of the regulatory review process may decrease sponsor hesitancy to develop innovative combination products. There has been concern that the advantages of being first-to-market with a new therapeutic class are more

TABLE 7.4
Recommended Pharmacokinetic/Toxicokinetic and Drug Metabolism Activities for Combination Drug Products Containing Marketed Drugs and/or New Molecular Entities

Nonclinical Area	Recommended Activity
Pharmacokinetics	
Marketed drug	Evaluate available PK and TK database, and if necessary, conduct additional studies to further characterize the PK profile of the marketed drug.
NME	Conduct the necessary PK and TK studies to characterize the absorption, distribution, metabolism, and elimination profiles of the individual NME.
CDP with only marketed drugs	Provide TK support for the recommended toxicology studies (bridging and developmental toxicology) on the CDP, and if potential PK interactions between the components in a CDP are expected, design and conduct animal PK studies to evaluate and characterize these possible drug-drug interactions.
CPD with marketed drugs and one NME	Conduct recommended PK and TK studies on the NME alone, provide TK support for the recommended toxicology studies on the CDP, and if potential PK interactions between the components in a CDP are expected, design and conduct animal PK studies to evaluate and characterize these possible drug-drug interactions.
CPD with more than one NME	Conduct recommended PK and TK studies on each NME alone, provide TK support for the recommended toxicology studies on the CDP, and if potential PK interactions between the components in a CDP are expected, design and conduct animal PK studies to evaluate and characterize these possible drug-drug interactions.
Drug Metabolism	
Marketed drug	Evaluate available DM database (e.g., tissue distribution, protein binding, metabolism [including enzyme systems involved and potential for enzyme inhibition or induction], and routes and rates of elimination) on each marketed drug, and if necessary, conduct additional DM studies to further characterize the DM profile of a marketed drug to meet current regulatory standards.
NME	Conduct recommended DM studies on the NME alone to characterize how the NME is handled by the body.
CDP with only marketed drugs	Compare the available and sponsor-generated DM information on each marketed drug in the CDP, and if necessary, design and conduct DM studies on the CDP to determine if the DM profiles of the individual components are changed when administered in combination.
CPD with marketed drugs and one NME	Compare the DM information on the marketed drugs with the DM information generated by the sponsor on the NME, and if necessary, design and conduct DM studies on the CDP to determine if the DM profiles of the individual components are changed when administered in combination.
CPD with more than one NME	Compare the DM information generated by the sponsor on the NMEs alone (and on any marketed drugs), and if necessary, design and conduct DM studies on the CDP to determine if the DM profiles of the individual NMEs (and any marketed drugs) are changed when administered in combination.

TABLE 7.5
Recommended Nonclinical Safety or Toxicology Activities for Combination Drug Products Containing Marketed Drugs and/or New Molecular Entities

Nonclinical Area	Recommended Activity
General Toxicology	
Marketed drug	Evaluate available toxicology database to ensure sufficient data are available to meet current regulatory recommendations for acute and repeated dose toxicology, genotoxicity, carcinogenicity, reproductive and developmental toxicology, and local tolerance, and if necessary, conduct additional studies to further characterize the toxicology profiles of the marketed drug in rodent and non-rodent species.
NME	Conduct the recommended toxicology studies, which may be different for a NCE and a biological, to characterize the animal safety profiles of the individual NME in rodent and non-rodent species
CDP with only marketed drugs	After the toxicology database for each marketed drug in a CDP is complete, compare the toxicology profiles of each component in the CDP to determine if additional toxicology studies, other than the toxicology bridging study and a developmental toxicology study (discussed below), are necessary to evaluate and characterize a potential toxicological interaction between two or more of the components in a CDP.
CPD with marketed drugs and one NME	After the toxicology database for each marketed drug is complete and the recommended toxicology studies on the NME have been conducted, compare the toxicology profiles of each component in a CDP to determine if additional toxicology studies, other than the toxicology bridging study and a developmental toxicology study (discussed below), are necessary to characterize a potential toxicological interaction between the components in a CDP.
CPD with more than one NME	After the recommended toxicology studies on each NME in a CDP have been conducted and the toxicology database for any marketed drugs in the CDP is considered complete, compare the toxicology profiles of each component in a CDP to determine if additional toxicology assessments are needed to further characterize the toxicology profiles of components administered concurrently.
Recommended Bridging Toxicology Study	
CDP with only marketed drugs	Design and conduct the recommended toxicology bridging study in the most relevant animal species using multiple dose levels of the CDP, the projected clinical route and frequency of dosing for each marketed drug, and dose groups for each of the marketed drugs alone.
CPD with marketed drugs and one NME	Design and conduct the recommended toxicology bridging study in the most relevant animal species using multiple dose levels of the CDP, the projected clinical route and frequency of dosing for the NME and for each marketed drug, and dose groups for the NME alone and for each of the marketed drugs alone.

(Continued)

TABLE 7.5 (CONTINUED)
Recommended Nonclinical Safety or Toxicology Activities for Combination Drug Products Containing Marketed Drugs and/or New Molecular Entities

Nonclinical Area	Recommended Activity
CPD with more than one NME	Design and conduct the recommended toxicology bridging study in the most relevant animal species using multiple dose levels of the CDP, the projected clinical route and frequency of dosing for each NME in the CDP, and, if included, for each marketed drug in the CDP, and dose groups for each of the NMEs alone and, if included, in the CDP, for each of the marketed drugs alone.
Recommended Developmental Toxicology Study	
CDP with only marketed drugs	Unless one or more of the marketed drugs in a CDP is known to have significant risk for developmental toxicity, design and conduct (at the appropriate time in the CDP development program) the recommended embryo-fetal developmental toxicology study on a CDP.
CPD with marketed drugs and one NME	Unless one or more of the marketed drugs in a CDP is known to have significant risk for developmental toxicity or if the NME alone is shown to cause reproductive and developmental toxic effects in rodents or non-rodents, design and conduct the recommended embryo-fetal developmental toxicology studies on a CDP.
CPD with more than one NME	If any of the NMEs alone in a CDP is shown to cause reproductive and developmental toxic effects in rodents or non-rodents and unless any marketed drugs, if included, in the CDP is known to have significant risk for developmental toxicity, design and conduct the recommended embryo-fetal developmental toxicology studies on a CDP.

than offset by the complications and delays associated with being the test case as FDA determines the best approach to review the given product type. Even when efforts between reviewing divisions are coordinated, the need to involve multiple divisions increases FDA review time. Reviewing divisions have been less responsive to the mandated response time periods when they are not the primary reviewing division than when they are a consulting division, resulting in delays for a complete review of combination product data and plans.

One challenge for clinical development of a combination product is the lack of specific guidance. The OCP has issued guidance documents on preclinical and chemistry, manufacturing, and controls (CMC) considerations. For clinical programs, however, the amount of data and approach to development depend heavily on the reviewers' previous experience with the components as well as any unique issues raised by the combination. A recent series of guidances covering drug-eluting stents represents FDA's first attempt to provide global and detailed guidance on an important class of combination products that are widely used in the general population.[28,29] These guidances cover suggested development pathways for these specific products, safety support for drug substance and combination product, finished product handling (e.g., sterilization, package integrity, and shelf life) and clinical assessment, and CMC information required.

The advent of nanotechnology-based products and delivery systems or vehicles present special challenges and are considered on a case-by-case basis by the applicable FDA reviewing division(s). Issues include absorption and ultra-small particle sizes for active ingredients/excipients that may act differently at the molecular receptor level than larger particles. There are not yet specific guidances available on the pharmacological or clinical development of such products.

One common factor among the more successful (and rapid) combination product approvals is early and frequent communication among developers, FDA reviewers, and, as appropriate, the OCP. Organized teleconferences with the decision-makers in all reviewing divisions (and with all sponsors, when there are more than one) have been especially beneficial in establishing consensus, avoiding weeks and even months of email and letter "loops." Complete reviewer buy-in on the development program (e.g., appropriateness of surrogate endpoints, number of studies, and indications) and individual trial issues (e.g., study design, sample size, and endpoints) can avoid extended review periods and unnecessary duplication of research efforts. New combination product developers can especially benefit from obtaining FDA guidance on how to approach unique issues such as techniques to measure drug levels in tissues not typically accessible (for targeted therapies) or techniques to evaluate drug-device interactions.[27] Table 7.6 and Table 7.7 provide examples of common drug/delivery system and drug/active device combination products, respectively, with the advantages and challenges of each type.

While the safety and effectiveness of drug and biologic products are mainly intrinsic to the products themselves, the safety and effectiveness of devices often have a human-use component that should be examined in clinical development. For example, improper patient use of a metered dose inhaler can drastically reduce the efficacy of the inhaled drug. Studies of how patients operate the device in "realistic, stressful conditions" should be conducted early in development, as they will provide insight into possible device improvements prior to substantial evidence studies.[30]

TABLE 7.6
Examples of Drug-Delivery System Combinations

Examples	Advantages	Challenges
• Metered dose inhalers • Transdermal patches	• Streamlined regulatory process possible • Improved efficacy over previously approved delivery routes • Non-injection bioavailability for peptides and proteins	• Altered pharmacodynamic effects • Route-dependent metabolic profile • Inherent delivery site sensitivity • Safety of excipients in novel delivery routes • Possible immune system reactions • Bridging effects specific to human physiology • Formulation changes

TABLE 7.7
Examples of Drug-Active Device Combination

Examples	Advantages	Challenges
• Coronary drug-eluting stents • Antibiotic bone cement	• Increased efficacy over device alone • Localized drug can reduce systemic adverse effects	• Often requires request for designation process • Localization of drug can alter safety profile

CDER has a long history of reviewing products that combine two or more active drug ingredients at fixed doses. As both active drugs carry the risk of side effects, CDER requires developers to demonstrate in clinical studies that each component contributes to the overall efficacy. CDER will often want to see data to justify each component in a drug/device combination, which can increase the size, cost, and complexity of a clinical program. Instead, it is often advantageous to have the product regulated as a device and assigned to CDRH, which reviews device systems that are often complex and reviews the safety and efficacy of the product as a whole.[30, 31]

Table 7.8 provides a summary of advantages and challenges for drug/in vitro diagnostic combination products with two examples of such products.

While all FDA centers require that products be demonstrated to be both safe and effective to be approved, the amount of clinical support each requires is different. CDRH requires a single high-quality substantial evidence study for approval while, in most cases, CDER requires two independently conducted substantial evidence trials. In situations where a disease is serious or life-threatening and there is an unmet medical need (i.e., no product approved in the US for the specific indication), or the prevalence of the disease is very low, CDER may approve a product with one well-controlled substantial evidence trial plus strong supporting information from earlier, smaller trials. A typical NDA approved by CDER may contain approximately 60 clinical studies (including dose-finding, pharmacokinetic, drug-interaction, and

TABLE 7.8
Drug-In Vitro Diagnostic Combination

Examples	Advantages	Challenges
Helicobacter pylori detection test	• May permit approval for products without limited general efficacy • May expand use of pharmaceuticals with significant safety concerns • Smaller clinical studies possible	• Biomarker development and validation • Possible limiting of target population • Increased complexity of development program

special population studies), while a PMA approved by CDRH may only contain two or three studies. Streamlined approvals are critical for devices, which can become obsolete within 6 to 12 months. For biologics, at least one well-controlled substantial evidence clinical trial must be included in a BLA; in practice, two or three such trials may be necessary, depending on the indication for the product and the weight of efficacy and safety data.

Of course, the amount of clinical trial data required to support approval of a multi-active combination product (under either a PMA or an NDA) depends on whether device or drug components of the product have been previously approved. If the drug to be used is unapproved, or is an analog of an approved drug, it is considered a new molecular entity by both CDER and CDRH, and manufacturers will need to provide information in the IDE equivalent to the content of a Phase 1 IND for CDER. The eventual NDA for a CDER-regulated combination product must similarly include results of a robust clinical program that establishes substantial evidence of safety and effectiveness for the new product.

The number of combination products developed in the future will depend heavily on the success of products approved for marketing. Of concern for developers are the delays involved in being an innovator product, where the regulatory approval path is not well established. The approval process may remain challenging as regulators require the developers of follow-on products to consider safety issues of the innovator products in their development programs.

Even with the FDA's efforts to make the approval process consistent and transparent, the developers of combination products will have to consider the expectations and review culture of the different divisions in their positioning of the primary mode of action for the product. Developers should maximize opportunities for meetings (and/or teleconferences) that include both the main reviewing division and the consulting division(s). These meetings are critical in developing a streamlined clinical program that will address the needs of all reviewers, and in addressing unique issues that arise from the interaction of the combination product components.

CLINICAL INVESTIGATION AND PREMARKET REVIEW REQUIREMENTS FOR DRUGS AND MEDICAL DEVICES

The FDA has not established clinical investigation and premarket review requirements specifically applicable to typical combination products. Therefore, combination products are regulated using the requirements that have been established for the components of these products. These requirements are universally applied with regard to their overall intent to assure that clinical investigations are designed to be scientifically sound and adequately protect human subjects, and that products approved or cleared for marketing are safe and effective for their intended use. There are differences with regard to specific requirements, however, which can have an impact on the marketing application(s) that are submitted for review. Please see Table 7.9 for a description of these issues.

TABLE 7.9
Clinical Investigation and Premarket Submission Requirements for Drugs, Biological Products, and Medical Devices

Product	Clinical Investigation Requirements[1]	Premarket Review Requirements
Drug	IND (21 CFR Part 312)	NDA
Generic drug	IND[2]	ANDA
Biological product	IND	BLA[3]
Medical device		
Class I	IDE[4]	510(k)[5]
Class II	IDE[4]	510(k)[5]
Class III	IDE[4]	510(k)[6] or PMA

1. The clinical investigation of drugs, therapeutic biological products, and medical devices must be conducted in compliance with the requirements for institutional review set forth in 21 CFR Part 56 and with the requirements for informed consent set forth in 21 CFR Part 50.
2. Clinical data are normally not required to support an ANDA; however, an IND is required to conduct clinical investigations of any unapproved new drug.
3. A BLA is required for any biological drug product regardless of the FDA center with jurisdiction over the product.
4. Clinical data are not required to support most 510(k) premarket notification submissions. When clinical studies are conducted, an IDE approved by an institutional review board (IRB) is required unless exempted. If the study involves a significant risk device, the IDE must also be approved by the FDA.
5. For device classifications not exempted by regulation from premarket notification requirements.
6. Class III "preamendment" device classifications for which FDA has not established an effective date for the submission of PMA applications may be cleared for marketing via the 510(k) substantial equivalence process.

The level of premarket review required for devices is risk-based. Devices are classified by regulation into one of three regulatory classes: I, II, or III. Most Class I and certain Class II devices are exempt from premarket approval or clearance requirements. Most Class II devices require a 510(k) premarket notification to the CDRH. The 510(k) submission must contain information and data establishing that the device is substantially equivalent to one or more identified "predicate devices" that may be legally marketed in the US for the same intended use. In most instances, clinical data are not required to support the 510(k) premarket notification submission. The device may not be marketed until FDA responds to the 510(k) submission that the device is cleared for marketing.

If FDA responds to the 510(k) submission that the device is not substantially equivalent to a predicate device, the device is automatically considered a Class III device requiring an approved PMA application before it can be marketed.[32]

The FDA has recently issued a guidance on device classification that has an impact on the mechanisms and requirements for reviewing drug/device and biologics/device submissions.[33] The purpose of this guidance is to establish procedures for submitting, reviewing, and responding to requests for information regarding the class in which a device has been classified or the requirements applicable to a device on the FD&C Act that are submitted in accordance with section 513 (g), which deals with such information requests for classification by the Agency.[33]

To market most Class III devices, a PMA application must be approved by FDA. The PMA application is similar to an NDA or BLA in that it must contain sufficient valid scientific evidence from nonclinical and clinical studies to show that the product is safe and effective for its intended use(s). Devices that were on the market prior the date of enactment of the medical device amendments to the FD&C Act (May 28, 1976) are called preamendment devices. Sponsors of Class III preamendment devices require a PMA application only after FDA publishes a final regulation calling for a PMA for the types of devices covered by that regulation, and establishes an effective date for submission of the PMA application. Devices that are of the same type as a Class III preamendment device for which FDA has not yet established an effective date for a PMA submission require 510(k) clearance prior to marketing. The majority of devices that are subject to premarket requirements (Class I reserved, Class II, and Class III) are cleared for marketing via the 510(k) substantial equivalence process.[31]

The implications of a jurisdictional assignment for a combination product with respect to its primary mode of action can be significant.[34] For example, user fees for medical device marketing applications are substantially lower than those for products regulated as drugs. Expected time for FDA to review marketing applications is also a factor. The target review time for a device 510(k) premarket notification, for example, can be as little as 30 days, while target review times for an NDA or BLA can be six to ten months. Also, companies that are used to dealing with a certain product type (e.g., devices) may not be familiar with the regulatory requirements and options applicable to other types of products (e.g., device companies may not be familiar with IND requirements for drug products or Fast Track. Rare Pediatric Disease, or Breakthrough Therapy designation for qualifying drug products) that may be applicable to the combination product based on its jurisdictional assignment. Table 7.10 presents examples of jurisdictional determinations for combination products.

The electronic (eCTD) marketing application requirements for NDAs are also detailed, complex, and expensive, and developers of combination products must take this into account. The sponsor's strategic regulatory and product development approach may be to seek marketing approval or clearance of a component of their combination product as well as for the combination product as a whole by submitting separate marketing applications. The sponsor of a drug-device combination regulated by CDER may, for instance, in addition to seeking approval of an NDA for the drug-device combination, intend to submit a separate PMA or 510(k) application to CDRH for approval or clearance, respectively, of the device component as a separate or stand-alone product. In this case, the sponsor would likely wish to

TABLE 7.10
Examples of Jurisdictional Determination in Accordance with 21 CFR Part 3

Example 1

Product: Spinal fusion device coated with a therapeutic protein intended to treat degenerative disc disease. A spinal fusion cage soaked in a solution of a therapeutic protein to coat the inside surfaces of the device. In this hypothetical example, the fusion cage, a permanent implant, maintains the spacing and stabilizes the diseased region of the spine, while the protein is used to encourage the formation of bone within the fusion cage to further stabilize this portion of the spine as well as the cage itself.

Modes of Action (MOA)/Primary Mode of Action (PMOA): The PMOA is attributable to the device component's MOA to mechanically maintain the intervertebral spacing and stabilize the diseased region of the spine, while the therapeutic protein's MOA to encourage bone formation within and around the cage plays a secondary role. In this hypothetical example, the therapeutic protein does not have the mechanical properties necessary to maintain the spacing and stabilize the spine if used alone. Furthermore, clinically successful spinal fusion (pain reduction and stability of the spine) can be achieved even in the absence of bone growth within the cage. The device component provides the most important therapeutic action of the combination. It is unnecessary to proceed to the assignment algorithm because it is possible to determine which MOA provides the most important therapeutic action of this combination product.

Assignment Algorithm Criteria: not applicable.

Jurisdictional Assignment: CDRH.

Example 2

Product: Chemotherapeutic drug and monoclonal antibody for targeted cancer treatment. The monoclonal antibody is intended to improve the drug's effectiveness by directly targeting the drug to receptors on cancer tumor cells.

Modes of Action (MOA)/Primary Mode of Action (PMOA): The PMOA is attributable to the drug component's MOA (cytotoxic action on cancer cells), while the biological product component's MOA to target the drug to the receptors on the cancer cells enhances the efficacy of the drug. The drug component provides the most important therapeutic action of the combination. It is unnecessary to proceed to the assignment algorithm because it is possible to determine which MOA provides the most important therapeutic action of this combination product. Note that in June 2003, FDA transferred to CDER the regulation of certain therapeutic biological products that had been regulated by CBER, including monoclonal antibodies. Although CDER now has regulatory responsibility over both the chemotherapeutic drug and monoclonal antibody described in this hypothetical example, this example is provided for illustrative purposes.

Assignment Algorithm Criteria: not applicable.

Jurisdictional Assignment: CDER.

Example 3

Product: Scaffold seeded with autologous cells for organ replacement. The hypothetical product has the shape of the target organ, and the autologous cells are intended to allow the product to ultimately function like the target organ in the patient.

Modes of Action (MOA)/Primary Mode of Action (PMOA): The PMOA is attributable to the biological product component's MOA to help form new organ tissue that will ultimately function like the native organ. The device component's MOA to provide a scaffold upon which the new tissue will form is secondary. Although the scaffold is necessary to create the new tissue and provide the necessary shape, the creation of a functioning organ is primarily dependent upon the role of the cells to provide the tissue organization and muscular layer needed to function like the native organ. The biological product component provides the most important therapeutic action of the combination. It is unnecessary to proceed to the assignment algorithm because it is possible to determine which MOA provides the most important therapeutic action of this combination product.

Assignment Algorithm Criteria: not applicable.

Jurisdictional Assignment: CBER.

(Continued)

TABLE 7.10 (CONTINUED)
Examples of Jurisdictional Determination in Accordance with 21 CFR Part 3

Example 4

Product: Menstrual tampon impregnated with genetically modified bacteria. The hypothetical product is intended for use throughout menstruation both in the collection of menstrual fluid and to treat and/or prevent recurrence of bacterial vaginosis.

Modes of Action (MOA)/Primary Mode of Action (PMOA): The product has two MOAs – the action of the biological product component to act upon the vaginal mucus membrane to produce antimicrobial factors that will control opportunistic pathogens, and the device component's action to collect menstrual fluid. Both actions are independent and neither appears to be subordinate to the other so it is necessary to apply the assignment algorithm.

Assignment Algorithm Criteria: 1) Is there a center that regulates other combination products that present similar questions of safety and effectiveness with regard to the combination product as a whole? CDRH regulates tampons, and CBER regulates bacterial products and genetically modified cells. In this example, no combination product intended to collect menstrual fluid and to treat and/or prevent recurrence of bacterial vaginosis through the actions of a genetically modified organism has previously been reviewed by the Agency. Although both CDRH and CBER regulate products that raise similar safety and effectiveness questions with regard to the product components, neither center regulates combination products that present similar safety and effectiveness questions with regard to the product as a whole so it is necessary to apply the second criterion.

2) Which center has the most expertise related to the most significant safety and effectiveness questions presented by the combination product? Because there is no center that regulates combination products that present similar safety and effectiveness issues with regard to the product as a whole, the FDA would consider which center has the most expertise related to the most significant safety and effectiveness questions presented by the product. In this case, the menstrual tampon component presents generally routine safety and effectiveness questions, similar to those of other menstrual tampons. In contrast, the biological product component raises more significant safety and effectiveness questions, such as those related to bacterial strain selection and dose; bacterial purity, potency, and metabolic activity, including the impact of genetic modifications; and bacterial adherence potential, microbial strain interactions, and constitutive production of ancillary antimicrobial substances.

Jurisdictional Assignment: CBER.

Example 5

Product: Interferon and ribavirin combination therapy. The product is intended for use in the treatment of chronic hepatitis C. Interferon is approved under the licensing provisions of the Public Health Service Act as a single-entity product for treatment of chronic hepatitis C. Clinical studies show that ribavirin when used alone to treat chronic hepatitis C can improve liver function, but most patients relapse with treatment of ribavirin alone. Data show that when used in conjunction with interferon, ribavirin produces a more efficacious response than when interferon is used alone to treat chronic hepatitis C. The drug and biological product components may be copackaged or are provided separately but cross-labeled for use together.

Modes of Action (MOA)/Primary Mode of Action (PMOA): The PMOA is attributable to the biological product component's MOA to treat chronic hepatitis C, which produces a dose-dependent decline in hepatitis C virus ribonucleic acid (RNA) titers, while the drug component's MOA is to enhance the efficacy of the biological product. Note that interferons are now reviewed in CDER following the transfer of therapeutic biological products to CDER in 2003.

Assignment Algorithm Criteria: not applicable.

Jurisdictional Assignment: CDER.

(Continued)

TABLE 7.10 (CONTINUED)
Examples of Jurisdictional Determination in Accordance with 21 CFR Part 3

Example 6

Product: Implantable device with local chemotherapeutic drug. Embolization device coated with a chemotherapeutic agent intended to treat hypervascularized tumors.

Modes of Action (MOA)/Primary Mode of Action (PMOA): In this hypothetical example, the embolization device is a permanent implant, while the drug component is a short-term acting chemotherapeutic. The PMOA is attributable to the device component's MOA to physically occlude the blood supply to the tumor site through embolization, while the drug component's MOA plays a subordinate role in causing apoptosis in any remaining proliferating tumor cells. Data indicate that the effectiveness of the embolization device alone for the stated indication is much greater than the effectiveness of the drug component when delivered directly to the tumor site without use of the embolization agent. It is unnecessary to proceed to the assignment algorithm because it is possible to determine which MOA provides the most important therapeutic action of this combination product. In this hypothetical example, the PMOA was attributable to the device component; however, such a product used for another indication, or with another drug, could have a drug PMOA depending on the relative effectiveness of the drug and device components in providing the most important therapeutic action for the new use.

Assignment Algorithm Criteria: not applicable.

Jurisdictional Assignment: CDRH.

Example 7

Product: Vertebroplasty implant with extended-release analgesic. This hypothetical product is intended to provide spinal stabilization in patients with spinal bone metastases who also require palliative relief of pain.

Modes of Action (MOA)/Primary Mode of Action (PMOA): The product has two MOAs – the device action to stabilize the fractured spinal vertebral body bone and the drug action to provide for extended analgesic delivery as an alternative to oral medication in patients expected to continue to require long-term pain management despite the stabilization implant. Both actions of the product are independent, and neither is clearly subordinate to the other so it is necessary to apply the assignment algorithm.

Assignment Algorithm Criteria: 1) Is there a center that regulates other combination products that present similar questions of safety and effectiveness with regard to the combination product as a whole? CDRH regulates vertebroplasty implants and CDER regulates analgesic drug products. No product combining a vertebroplasty implant and an extended-release analgesic has yet been submitted to FDA for review, therefore neither center regulates combination products that present similar safety and effectiveness questions with regard to the product as a whole and it is necessary to apply the second criterion of the algorithm.

2) Which center has the most expertise related to the most significant safety and effectiveness questions presented by the combination product? Although important safety and effectiveness questions are presented by this new route of administration of an analgesic and its extended release from the device that would need to be addressed, in this hypothetical example, the most significant safety and effectiveness questions associated with the combination product as a whole are related to the mechanical strength, wear, and clinical performance of the vertebroplasty implant.

Jurisdictional Assignment: CDRH.

assure that all communications with the FDA regarding both development programs are appropriately coordinated between reviewing centers to assure that the safety and effectiveness issues associated with the device component of the combination and the stand-alone device can be addressed as efficiently as possible, and that the required timing of each marketing application is understood. In such instances, communication with the FDA should be initiated as early as possible in the development program so that the FDA can provide guidance on how best to proceed.

The number of marketing applications that the FDA may require or that the sponsor would like to submit should be included among the topics covered in discussions with the FDA early in the development process. Sponsors should keep in mind that the regulation of combination products is evolving as the FDA deals with new issues presented by innovative combination products. Sponsors should review current regulations and guidance documents, recent jurisdictional determinations, proposed rules, and other information the FDA makes available on its website before contacting the FDA with questions or to discuss specific issues that may be applicable to the sponsor's combination product development program.

A 2016 FDA guidance provides guidance to industry and FDA staff on the underlying principles of human factors (HF) studies during the development of combination products as defined under 21 CFR Part 3.[35] It describes Agency recommendations regarding HF information in a combination product investigational or marketing application and clarifies the different types of HF studies; the recommended timing and sequencing of HF studies; and how HF studies are part of the process to maximize the likelihood that the combination product user interface is safe and effective for use by the intended users, uses, and environments. In addition, the guidance describes how HF studies relate to other clinical studies. The guidance also provides process considerations for HF information in investigational or marketing applications to promote development and timely review of safe and effective combination products.

For medical devices, the use of human factors and usability engineering (e.g., applying the knowledge of human behavior, abilities, and limitations to the design of a medical device) plays a key role in maximizing the likelihood that the device will be safe and effective for use by the intended users, for the intended uses, and for the intended use environments. Under the medical device design control requirements described in 21 CFR 820.30, design validation must include a risk analysis where appropriate. As part of the risk analysis, device manufacturers should identify and analyze potential use-related hazards, including lessons learned from reported errors with similar products, and as appropriate, incorporate and validate design features that mitigate or eliminate these hazards. This assessment informs the device design development to eliminate or minimize use errors that could cause harm or compromise medical treatment.

REGULATORY STRATEGY CONSIDERATIONS FOR CHEMISTRY, MANUFACTURING, AND CONTROLS

Compliant combination product development, leading to a successful NDA, BLA, or PMA, is designed to meet the requirements of a product's development throughout its lifecycle by incorporating into the planning and execution of work all requisite elements of traceability to assure the validity and retrievability of information. The building blocks of this approach are planning, performing, recording, reporting, monitoring, and change control. Since both current good manufacturing practice (CGMP) and Quality System issues are involved in the typical combination product manufacturing controls arena, a careful approach to this area will lead to a more successful regulatory outcome for an otherwise safe and effective combination product during the marketing application review cycle.[36–38]

Regulatory-compliant product development is built upon the basic tenet of pharmaceutical development that product quality is by design. Utilizing good documentation practices, a sponsor can demonstrate an understanding that product and process knowledge will be gained throughout the lifecycle of a product. Regulatory-compliant product development activities are designed to gain insight and understanding of the effects that sources of variability will introduce to product quality. These insights and understandings form the basis of quality risk management programs. The information and process knowledge garnered through development activities and manufacturing experiences form the basis of specifications and manufacturing controls that assure that product is consistently manufactured to established standards of quality. If properly designed and implemented, a strategic CMC plan with the appropriate tactical execution will assist the CDP sponsor in meeting the requirements of parallel drug and device manufacturing systems.[39, 40]

The FDA has acknowledged that drug and device manufacturing regulations overlap, are generally similar, and have the same objectives, i.e., to allow the manufacture of safe and effective products. For CDPs produced as single entities or packaged together, all applicable CGMP regulations are relevant. Table 7.11 provides a

TABLE 7.11
Key Current Good Manufacturing Practice Provisions to Consider during and after Joining Together Co-Packaged and Single-Entity Combination Products

If the Operating Manufacturing Control System is Part 820 (QS Regulation)*:
§ 211.84, testing and approval or rejection of components, drug product containers, and closures;
§ 211.103, calculation of yield;
§ 211.137, expiration dating;
§ 211.165, testing and release for distribution;
§ 211.166, stability testing;
§ 211.167, special testing requirements;
§ 211.170, reserve samples.

If the Operating Manufacturing Control System is Part 210/211 (CGMP Regulation)*:
§ 820.30, design controls;
§ 820.50, purchasing controls;
§ 820.100, corrective and preventative actions.

* Including all subsections, as appropriate.

delineation of key CGMP provisions to consider for co-packaged and single entity combination products involving small molecule or biological drugs and medical devices.

Components of a CDP are identifiable and assignable to specific FDA centers, CFR sections, and applicable guidelines. In developing an approach to regulatory compliance, sponsors are well served by understanding that the different centers have varying regulatory approaches. CDRH is most concerned with design controls, documentation supporting the administration of quality systems during development and manufacturing, and the application of ongoing risk assessments. Conversely, CDER is more focused on the characterization and control of changes to CDP development that may potentially affect the quality, strength, purity, or potency of the product. A primary challenge for a CDP sponsor is to understand both the divergent and convergent perspectives of the centers on the regulatory processes.

CDP developers and manufacturers must appropriately focus attention to provide FDA with the necessary CMC information. The organization of this information may be broadly categorized into the two primary areas of drug and device, keeping in mind that the effects of the drug on the device performance and the device's potential impact on the impurity profile, safety, and efficacy of the drug will be of primary importance to the regulatory review process. The principles of Quality by Design (QBD) are critical for all medical product types.

Effective regulatory submissions for CDPs require that the sponsor present information that details the evolution of process and product knowledge from the perspective of CMC development. The approaches taken should be integrated at the foundational level of planning, performing, recording, reporting, and monitoring of the CDP development activities to communicate to CMC reviewers that the application of well documented change control has resulted in the identification, understanding, and control of sources of variability. The sponsor should demonstrate that the resultant continual improvements in the design and manufacturing of the CDP are associated with reductions in manufacturing and performance variability, which in turn underlie the sponsor's risk management program.

The framework for the CMC section of regulatory submissions remains constant throughout the lifecycle of a CDP. The information required is influenced by the stage of the product's development and risk-based considerations pertaining to the safety and effectiveness of its use.[36, 37, 40, 41] If submitting an NDA as the end result of a CDP development program, the sponsor must consider the incorporation of all of these elements in planning for the marketing submission.

The FDA has issued a regulation on the CGMP requirements applicable to combination products.[37] This rule is intended to promote public health by clarifying which CGMP requirements apply when drugs, devices, and biological products are combined to create combination products. In addition, the rule sets forth a transparent and streamlined regulatory framework for firms to use when demonstrating compliance with CGMP requirements for "single-entity" and "co-packaged" combination products.

According to the FDA, the constituent parts of a combination product retain their regulatory status (as a drug or device, for example) after they are combined.

Accordingly, the CGMP requirements that apply to each of the constituent parts continue to apply when they are combined to make combination products. The new regulations, in part, comprise the following:

§ 4.3 What Current Good Manufacturing Practice Requirements Apply to a Combination Product?

A sponsor who manufactures a combination product must follow the requirements listed in this section as follows:

(a) The CGMP requirements in parts 210 and 211 of this chapter apply to a combination product that includes a drug constituent part;
(b) The CGMP requirements in part 820 of this chapter apply to a combination product that includes a device constituent part;
(c) The CGMP requirements among the requirements (including standards) for biological products in parts 600 through 680 of this chapter apply to a combination product that includes a biological product constituent part to which those requirements would apply if that constituent part were not part of a combination product; and
(d) The current good tissue practice requirements including donor eligibility requirements for human cell and tissue products in part 1271 of this chapter apply to a combination product that includes a human cell and tissue product.

Section 4.4, on how sponsor can comply with these CGMP requirements for a co-packaged or single-entity combination product, includes the following points.

For single entity or co-packaged combination products, compliance with all applicable CGMP requirements for the combination product shall be achieved through the design and implementation of a CGMP operating system that is demonstrated to comply with the specifics of each set of CGMP regulations listed under § 4.3 as they apply to each constituent part included in the combination product or the following requirements apply:

- If the combination product includes a device constituent part and a drug constituent part, and the CGMP operating system has been shown to comply with the drug CGMPs, the following provisions of the Quality System (QS) regulation must also be shown to have been satisfied: management responsibility, design controls, purchasing controls, corrective and preventive action, installation, and servicing. Upon demonstration that these requirements have been satisfied, no additional showing of compliance with respect to the QS regulation need be made.
- If the combination product includes a device constituent part and a drug constituent part, and the CGMP operating system has been shown to comply with the QS regulation, the following provisions of the drug CGMPs

must also be shown to have been satisfied: Testing and approval or rejection of components, drug product containers, and closures; calculation of yield; tamper-evident packaging requirements for over-the-counter (OTC) human drug products; expiration dating; testing and release for distribution; stability testing; special testing requirements; and reserve samples. Upon demonstration that these requirements have been satisfied, no additional showing of compliance with respect to the drug CGMPs need be made.

- In addition to being shown to comply with the other applicable manufacturing requirements listed under § 4.3, if the combination product includes a biological product constituent part, the CGMP operating system must also be shown to implement and comply with all manufacturing requirements identified under § 4.3(c) that would apply to that biological product if that constituent part were not part of a combination product.
- In addition to being shown to comply with the other applicable CGMP requirements listed under § 4.3, if the combination product includes a human cell and tissue product, the CGMP operating system must also be shown to implement and comply with all current good tissue practice requirements identified under § 4.3(d) that would apply to that human cell and tissue product if it were not part of a combination product.

A 2017 guidance[40] describes and explains the final rule on current good manufacturing practice (CGMP) requirements for combination products that FDA issued on January 22, 2013.[41]

Key clarifications provided in the guidance include:

- The constituent parts of a combination product retain their regulatory status (as a drug or device, for example) after they are combined. CGMP requirements that apply to each of the constituent parts apply to the combination product they constitute.
- The final rule on CGMP requirements for combination products applies to all combination products. The CGMP requirements for constituent parts of cross-labeled combination products that are entirely manufactured at separate facilities are the same as those that would apply if these constituent parts were not part of a combination product. With regard to cross-labeled combination products, the CGMP obligations applicable to the drugs, devices, or biological products also apply to these types of articles when they are constituent parts of cross-labeled combination products.
- For single-entity combination products and co-packaged combination products, part 4 identifies two ways to demonstrate compliance with CGMP requirements. Under the first option, manufacturers demonstrate compliance with all CGMP regulations applicable to each of the constituent parts included in the combination product. Under the second option, manufacturers implement a streamlined approach for combination products that include both a drug and a device by demonstrating compliance with either the drug CGMPs (21 CFR parts 210 and 211) or the device Quality System

(QS) regulation (21 CFR part 820) and also demonstrating compliance with specified provisions from the other of these two sets of CGMP requirements. In addition, for a combination product that includes a biological product, the manufacturer must demonstrate compliance with the CGMP requirements specific to biological products in parts 600 through 680 (21 CFR parts 600 through 680). For a combination product that includes any HCT/P, the manufacturer must demonstrate compliance with the regulations in part 1271 (21 CFR part 1271) – including the current good tissue practice (CGTP) requirements and donor eligibility requirements.

- If a facility manufactures a drug or device that is not a constituent part of a combination product and also manufactures a combination product, the CGMP requirements for the independently marketed drug or device do not change (compliance with parts 210/211 or part 820, respectively). Accordingly, if a facility manufactures an independently marketed device that is subject to part 820 and a combination product, it cannot manufacture the device under a drug CGMP-based streamlined operating system, even if it is used for the combination product, because this operating system only includes limited, specified provisions from the QS regulation. However, both the device and the combination product may be manufactured under a device QS regulation-based streamlined operating system, with the device that is not part of a combination product then being subject only to part 820.
- A facility that manufactures only device components, including device components used in a combination product, is not made subject to the device QS regulation by part 4. Similarly, a facility that manufactures solely active pharmaceutical ingredients (APIs) is not subject to part 211, though it is subject to statutory CGMP requirements under Section 501 of the FD&C Act (21 U.S.C. 351). Part 4 does not make such a drug component manufacturing facility subject to part 211 if those components are for use in a combination product. However, a facility that manufacturers a drug-device combination product formed from components is subject to parts 211 and 820, and must comply with these two sets of regulations as provided in part 4.
- The terms "constituent part" in part 4 and "component" in the CGMP regulations serve different regulatory purposes. The use of the term "constituent part" in part 4 refers to drugs, devices, and biological products included in combination products and does not alter the meaning of the term "component" or alter whether the regulations listed in 21 CFR 4.3 apply to component manufacturers.
- The Agency draws a distinction between mere drug containers and closures and containers and closures that are also devices. The essential difference is generally whether the article is designed to deliver the drug it contains or merely to hold it. If the article merely holds the drug, it is only subject to drug CGMPs as a container or closure. An article that holds or contains a drug, but also delivers it, may also be a device subject to the device QS regulation in addition to the requirements relating to drug containers and closures.

- A container closure system is the sum of packaging components that together contain and protect the drug product. This includes primary and secondary packaging components if the latter are intended to provide additional protection to the drug product. Elements of container closure systems that are device constituent parts include piston syringes, metered dose inhalers (MDIs), and containers for intravenously-administered (IV) fluids (such as IV bags containing saline or anti-coagulants), which both hold and deliver the drugs or biological products they contain. In the case of an IV container, the drug is delivered to the patient via the IV line.
- A facility that manufacturers a constituent part of a combination product or a complete combination product must be compliant with the CGMP requirements applicable to each manufacturing process that occurs at that specific facility. In addition, the combination product owner (the holder of the marketing authorization for the product) retains overall responsibility for the product, even if the owner is not directly engaged in its manufacture. Quality agreements and audits can be helpful in assuring compliance with applicable CGMP requirements.
- In addition to complying with drug CGMP and device QS regulation requirements as applicable in accordance with part 4, a manufacturer of a combination product that contains a biological product or an HCT/P must comply with the requirements that would apply to the biological product or HCT/P if it were not part of a combination product.
- A biological product regulated under section 351 of the PHS Act is also, by definition, a drug or a device. Accordingly, in addition to the requirements in parts 600 through 680, a biological product is always either subject to the drug CGMPs or subject to the device QS regulation, regardless of whether the biological product is a constituent part of a combination product. The CGMP requirements for biological products in parts 600 through 680 augment the drug CGMPs and device QS regulation to ensure adequate consideration of issues for biological products.
- The CGMP requirements for biological products in parts 600 through 680 must be satisfied if a combination product includes a biological product constituent part.
- An HCT/P that is not regulated solely under section 361 of the PHS Act and Part 1271 is also regulated as a drug, device, or biological product. The drug CGMPs, device QS regulation, and the requirements in parts 600 through 680 may apply to an HCT/P depending on whether the product is regulated as a drug, device, or biological product. CGMPs and the Current Good Tissue Practices (CGTPs) for HCT/Ps supplement one another and do not supersede each other unless the regulation specifically provides otherwise. In the event that a regulation in part 1271 is in conflict with a requirement in parts 210, 211, 600 through 680, or 820, the regulations more specifically applicable to the product in question will supersede the more general.

POSTAPPROVAL MODIFICATIONS FOR COMBINATION PRODUCTS

The FDA has issued a guidance on submission of requests for postapproval modifications to combination products approved under a BLA, NDA, or PMA.[42] Regardless of center assignment, in most instances the FDA may regulate the entire combination product under one type of marketing application (e.g., one BLA, NDA, or PMA). This one application would include all necessary information to support the approval of the combination product as a whole, including each of its constituent parts (drug, device, and/or biological product). If a change is made to any constituent part of the combination product that would have required a postmarket submission to FDA if the constituent part were a standalone product, then a postmarket submission is required for the combination product. In addition, a postmarket submission would also be required for the combination product if a change to any of the constituent parts would otherwise trigger the requirements associated with the application type used for approval of the combination product. In cases where the regulatory identity of the constituent part differs from the approved application type for the combination product, and a change is made that would require a postmarket submission to FDA, the requirement for submitting information about the change to the Agency is generally satisfied with one postmarket submission to the original application. The type of submission to provide for the change will depend on the type of application used to obtain approval of the combination product. For example, a change to the device constituent part of a combination product approved under an NDA should be reflected in the appropriate postmarket NDA submission and be submitted to that NDA.

In some cases, it may be easier to first identify the type of submission typically associated with the constituent part before determining what type of submission is required to the original application that was used for approval of the combination product.

Recent FDA guidance has clarified situations in which a postmarketing modification of one or more components of a drug-device or biologic-device can be effected for an approved or licensed combination product (e.g., a change in the bore size of a delivery device) in a way which has, if possible, minimal impact on the sponsor and FDA review staff, and preserves the safety and effectiveness of the product.[43] Since there are so many variables inherent in this discussion, sponsors will need to review and guidance and seek FDA consultation prior to effecting any of these changes.

RECENT FDA GUIDANCE ON COMBINATION PRODUCTS

The most recent guidance on combination products was issued in January 2022, in response to the requirements of the 2016 20th Century Cures Act that mandates that the FDA centers to apply the least burdensome approach for development of such products. It also makes clear and consistent its requirements for their development, review, and approval.[44] The guidance does not materially change any concepts of previous guidance as delineated in this chapter but rather incorporates by reference these documents and makes clear the Agency's overall approach to combination products and the INDs and marketing applications that are generated from their development programs.

REFERENCES

1. Siegel EB, ed. *Development and approval of combination products: A regulatory perspective.* Hoboken, NJ: John Wiley & Sons, Inc; 2008.
2. Federal Food, Drug and Cosmetic Act, 21 USC §321. Available at http://www.fda.gov/RegulatoryInformation/Legislation/FederalFoodDrugandCosmeticActFDCAct/default.htm
3. Public Health Service Act, 42 USC §201. Available at http://www.fda.gov/RegulatoryInformation/Legislation/ucm149256.htm
4. Drug and biological product consolidation. *Fed Reg* 2003;68(123):38067–38068. Available at http://www.fda.gov/OHRMS/DOCKETS/98fr/03-16242.pdf
5. FDA. Frequently asked questions about combination products. February 2011. Available at http://www.fda.gov/CombinationProducts/AboutCombinationProducts/ucm101496.htm
6. FDA. Requesting FDA feedback on combination products. December 2020. Available at https://www.fda.gov/media/133768/download
7. FDA. Principles of premarket pathways for combination products. Guidance for industry and FDA staff. February 2019. Available at https://www.fda.gov/media/119958/download
8. US Department of Health and Human Services, FDA, OCP, CBER, CDER, CDRH. Interpretation of the term "chemical action" in the definition of device under section 201(h) of the federal food, drug, and cosmetic act. Guidance for industry and FDA staff. June 2011. Available at http://www.fda.gov/RegulatoryInformation/Guidances/ucm259059.htm
9. US Department of Health and Human Services, FDA, OCP, CBER, CDER, CDRH. Guidance for industry and FDA staff: Classification of products as drugs and devices and additional product classification issues. June 2011. Available at http://www.fda.gov/RegulatoryInformation/Guidances/ucm258946.htm
10. US Department of Health and Human Services, FDA, OCP. Guidance for industry: How to write a request for designation (RFD). April 2011. Available at http://www.fda.gov/RegulatoryInformation/Guidances/ucm126053.htm
11. FDA. How to prepare a pre-request for designation (Pre-RFD). February 2018. Available at https://www.fda.gov/media/102706/download
12. US Department of Health and Human Services, FDA, CDRH, CBER. Acceptance and filing reviews for premarket approval applications (PMAs). Guidance for industry and FDA staff. December 31, 2012. Available at http://www.fda.gov/downloads/MedicalDevices/DeviceRegulationandGuidance/GuidanceDocuments/UCM313368.pdf
13. US Department of Health and Human Services, FDA, OCP. Guidance for industry and FDA staff: Application user fees for combination products. April 2005. Available at http://www.fda.gov/RegulatoryInformation/Guidances/ucm126017.htm
14. US Department of Health and Human Services, FDA, CDER, CBER. Guidance for industry: Fast track drug development programs - designation, development, and application review. January 2006. Available at http://www.fda.gov/downloads/Drugs/GuidanceComplianceRegulatoryInformation/Guidances/UCM079736.pdf
15. US Department of Health and Human Services, FDA, CDER, CBER. Guidance for industry: formal meetings with sponsors and applicants for PDUFA products. February 2000. Available at http://www.fda.gov/downloads/Drugs/GuidanceComplianceRegulatoryInformation/Guidances/UCM079744.pdf

16. US Department of Health and Human Services, FDA, CDRH. Early collaboration meetings under the FDA Modernization Act (FDAMA). Final guidance for industry and for CDRH staff. February 28, 2001. Available at http://www.fda.gov/MedicalDevices/DeviceRegulationandGuidance/GuidanceDocuments/ucm073604.htm
17. FDA. Postmarketing safety reporting for combination products. Guidance for industry and FDA staff. July 2019. Available at https://www.fda.gov/media/111788/download
18. FDA. Compliance policy for combination product postmarketing safety reporting immediately in effect. Guidance for industry and FDA staff. April 2019. Available at https://www.fda.gov/media/111795/download
19. FDA. Evaluation of devices used with regenerative medicine advanced therapies. Guidance for industry. February 2019. Available at https://www.fda.gov/media/120266/download
20. US Department of Health and Human Services, FDA, CDRH, CBER, CDER. Guidance for industry and FDA staff: In vitro companion diagnostic devices. July 14, 2011. Available at http://www.fda.gov/MedicalDevices/DeviceRegulationandGuidance/GuidanceDocuments/ucm262292.htm
21. US Department of Health and Human Services, FDA, OCP, CDRH, CDER. Guidance for industry: New contrast imaging indication considerations for devices and approved drug and biological products. December 2009. Available at http://www.fda.gov/downloads/RegulatoryInformation/Guidances/UCM195951.pdf
22. US Department of Health and Human Services, FDA, CDRH, CDER, CBER, OCP. Guidance for industry and FDA staff: Glass syringes for delivering drug and biological products: Technical information to supplement International Organization for Standardization (ISO) standard 11040-4. April, 2013. Available at http://www.fda.gov/downloads/RegulatoryInformation/Guidances/UCM346181.pdf
23. US Department of Health and Human Services, FDA, CDRH, CDER, CBER, OCP. Guidance for industry and FDA staff: Technical considerations for pen, jet, and related injectors intended for use with drugs and biological products. June 2013. Available at http://www.fda.gov/downloads/RegulatoryInformation/Guidances/UCM147095.pdf
24. FDA. Selection of the appropriate package type terms and recommendations for labeling injectable medical products packaged in multiple-dose, single-dose, and single-patient-use containers for human use. Guidance for industry. October 2018. Available at https://www.fda.gov/media/117883/download
25. Lakings DB. Nonclinical drug development: pharmacology, drug metabolism, and toxicology. In Guarino RA (ed.), *New drug approval process, accelerating global registration*. 4th ed. New York: Marcel Dekker; 2004:19–61.
26. Siegel EB, Lakings DB. 2006. Preclinical drug development handbook: Regulatory considerations. In Gad SC (ed.), *Pharmaceutical Development Handbook Series*. Weinheim: Wiley–VCH.
27. US Department of Health and Human Services, FDA, CDER: Guidance for industry: Nonclinical safety evaluations for drug or biologic combinations. March 2006. Available at http://www.fda.gov/downloads/Drugs/GuidanceComplianceRegulatoryInformation/Guidances/ucm079243.pdf
28. US Department of Health and Human Services, FDA, CDRH, CDER. Guidance for industry: Coronary drug-eluting stents-nonclinical and clinical studies. March 2008. Available at http://www.fda.gov/downloads/Drugs/GuidanceComplianceRegulatoryInformation/Guidances/UCM072193.pdf
29. US Department of Health and Human Services, FDA, CDRH, CDER. Guidance for industry: Coronary drug-eluting stents-nonclinical and clinical studies companion document. March 2008. Available at http://www.fda.gov/downloads/Drugs/GuidanceComplianceRegulatoryInformation/Guidances/UCM072196.pdf

30. US Department of Health and Human Services, FDA. Guidance for industry and FDA staff: Early development considerations for innovative combination products. September 2006. Available at http://www.fda.gov/RegulatoryInformation/Guidances/ucm126050.htm
31. US FDA. Regulation of combination products: FDA employee perspectives, October 2002. Available at www.fda.gov/oc/combination/perspectives.html
32. US Department of Health and Human Services, FDA, CDRH, Office of Device Evaluation. New section 513(f)(2) - evaluation of automatic Class III designation. Guidance for industry and CDRH staff. February 19, 1998. Available at http://www.fda.gov/MedicalDevices/DeviceRegulationandGuidance/GuidanceDocuments/ucm080195.htm
33. US Department of Health and Human Services, FDA, CDRH, CBER. Guidance for industry and FDA staff: FDA and industry procedures for section 513(g) requests for information under the federal food, drug, and cosmetic act. April 6, 2012. Available at http://www.fda.gov/MedicalDevices/DeviceRegulationandGuidance/GuidanceDocuments/ucm209841.htm
34. Definition of primary mode of action of a combination product. *Fed Reg* 2005;70(164):49848–49862. Available at http://www.fda.gov/ohrms/dockets/98fr/05-16527.htm.
35. FDA. Human factors studies and related clinical study considerations in combination product design and development. February 2016. Available at https://www.fda.gov/media/96018/download
36. US Department of Health and Human Services, FDA, CDER, CBER. Guidance for industry: Q8: Pharmaceutical development. May 2006. Available at http://www.fda.gov/downloads/RegulatoryInformation/Guidances/ucm128029.pdf
37. US Department of Health and Human Services, FDA, OCP. Guidance for industry and FDA: Current good manufacturing practice for combination products. September 2004. Available at http://www.fda.gov/downloads/Drugs/GuidanceComplianceRegulatoryInformation/Guidances/ucm070268.pdf
38. US Department of Health and Human Services, FDA, CDER, CBER, CVM, ORA. Guidance for industry: Quality systems approach to pharmaceutical CGMP regulations. September 2006. Available at http://www.fda.gov/downloads/Drugs/.../Guidances/UCM070337.pdf
39. US Food and Drug Administration. Manual of standard operating procedures and policies, general information - review, intercenter consultative/collaborative review process. Version 4, June 18, 2004. Available at http://www.fda.gov/downloads/RegulatoryInformation/Guidances/UCM126016.pdf
40. FDA. Guidance for industry and FDA staff: Current good manufacturing practice requirements for combination products. January 2017. Available at https://www.fda.gov/media/90425/download
41. Current good manufacturing practice (CGMP) requirements for combination products. *Fed Reg*, January 22, 2013;78:4307. Available at https://www.federalregister.gov/articles/2013/01/22/2013-01068/current-good-manufacturing-practice-requirements-for-combination-products
42. US Department of Health and Human Services, FDA, OCP, Office of Special Medical Programs, Office of the Commissioner. Guidance for industry and FDA staff: Submissions for postapproval modifications to a combination product approved under a BLA, NDA, or PMA. January 2013. Available at http://www.fda.gov/downloads/RegulatoryInformation/Guidances/UCM336230.pdf

43. FDA. Bridging for drug-device and biologic-device combination products. Guidance for industry. December 2019. Available at https://www.fda.gov/media/133676/download
44. FDA. Principles of premarket pathways for combination products. Guidance for industry and FDA staff. January 2022. Available at https://www.fda.gov/media/119958/download

8 FDA Expedited Pathways

Oliver V. Cácananta

OVERVIEW OF EXPEDITED PATHWAYS

The path from a groundbreaking discovery to the commercial approval of drugs and devices is long, often spanning years and consuming substantial financial and human resources, with no guarantee of success. Recognizing the need for a more streamlined process, particularly in the face of public health crises such as the thalidomide scandal of the 1950s, the AIDS epidemic in the 1980s, and today's global coronavirus pandemic, the United States Food and Drug Administration (FDA) implemented expedited pathways to facilitate easier access to critical, life-saving products.

The FDA's history of expedited pathways and product designations began in the 1980s, when the agency introduced the concept of fast-tracking the development and review of treatments for serious conditions with unmet medical needs. The passage of the Prescription Drug User Fee Act of 1992 (PDUFA) by Congress formalized this approach, setting performance goals for regulatory reviews. As a result, the median approval time for new drug applications has decreased significantly, from 27 months in 1993 to 12 months or less in 2015.[1] Note that expeditious approval of products does not compromise the rigorous scientific and clinical standards required for FDA approval.

Four key FDA programs offer incentives to accelerate the development and approval of new drugs and biologics. These programs emphasize early and active consultation with the FDA, which can lead to shorter review timelines. Applicants may qualify for one or more of the following programs:

1. Fast Track Designation: Speeds up the development and review of drugs that treat serious conditions and fill an unmet medical need.
2. Breakthrough Therapy Designation: Granted to drugs showing early evidence of substantial therapeutic effects and have the potential to provide significant benefits over existing treatments.
3. Priority Review Designation: Given to drugs that offer major advances in treatment or address conditions lacking adequate therapies.
4. Accelerated Approval: Allows the FDA to approve drugs based on surrogate endpoints reasonably likely to predict clinical benefits.

Table 8.1 summarizes the characteristics and requirements for each program.[2] In general, products must satisfy certain conceptual thresholds, including, addressing a serious condition and/or an unmet medical need, and where existing treatments

DOI: 10.1201/9781315117911-8

TABLE 8.1
Overview Comparison of FDA's Expedited Programs[a3]

	Fast Track	Breakthrough Therapy	Accelerated Approval	Priority Review
Established	1997	2012	1997	1992
Type	Designation	Designation	Approval Pathway	Designation
Timelines for submission request	With IND or after ideally no later than the pre-BLA/NDA meeting	With IND or after, ideally no later than the end-of-phase 2 (EOP2) meeting	Discuss the possibility of accelerated approval with the review division during development (e.g., supporting the use of the planned endpoint as a basis for approval and discussing the confirmatory trials)	With original BLA, NDA, or efficacy supplement
Qualifying criteria	• A drug that is intended to treat a serious condition **AND** nonclinical or clinical data demonstrate the potential to address unmet medical need; **OR** • A drug that has been designated as a qualified infectious disease product (QIDP)	A drug that is intended to treat a serious condition **AND** preliminary clinical evidence indicates that the drug may demonstrate substantial improvement on a clinically significant endpoint(s) over available therapies	A drug that treats a serious condition **AND** generally provides a meaningful advantage over available therapies **AND** demonstrates an effect on a surrogate endpoint that is reasonably likely to predict clinical benefit or on a clinical endpoint that can be measured earlier than irreversible morbidity or mortality (IMM) that is reasonably likely to predict an effect on IMM or other clinical benefit (i.e., an intermediate clinical endpoint)	• An application (original or efficacy supplement) for a drug that treats a serious condition **AND**, if approved, would provide a significant improvement in safety or effectiveness; **OR** • Any supplement that proposes a labeling change pursuant to a report on a pediatric study under 505A; **OR** • An application for a drug that has been designated as a QIDP **OR** • Any application or supplement for a drug submitted with a priority review voucher
Timelines for FDA response	Within 60 calendar days of receipt of the request	Within 60 calendar days of receipt of the request	Not specified	Within 60 calendar days of receipt of original BLA, NDA, or efficacy supplement

(Continued)

TABLE 8.1 (CONTINUED)
Overview Comparison of FDA's Expedited Programs[a][3]

	Fast Track	Breakthrough Therapy	Accelerated Approval	Priority Review
Features	• Actions to expedite development and review • Rolling review	• Intensive guidance on efficient drug development • Organizational commitment • Rolling review • Other actions to expedite review	Approval based on an effect on a surrogate endpoint or an intermediate clinical endpoint that is likely to predict a drug's clinical benefit	Shorter clock for review of marketing application (six months compared with ten-month standard review)
Additional considerations	FDA may rescind designation if it no longer meets the qualifying criteria for fast track; **OR the** sponsor may withdraw the fast track designation if the designation is no longer supported by emerging data or the drug development program is no longer being pursued	Designation may be rescinded if it no longer meets the qualifying criteria for breakthrough therapy; **OR the** sponsor may also withdraw the breakthrough therapy designation if the designation is no longer supported by emerging data or the drug development program is no longer being pursued.	• Promotional materials • Confirmatory trials to verify and describe the anticipated effect on IMM or other clinical benefit • Subject to expedited withdrawal	Designation will be assigned at the time of the original BLA, NDA, or efficacy supplement filing
Reference	Section 506(b) of the Food, Drug and Cosmetic Act (FD&C), as added by Section 112 of the FDA Modernization Act of 1997 (FDAMA), and amended by Section 901 of the FDA Safety and Innovation Act of 2012 (FDASIA)	Section 506(a) of the FD&C Act, as added by section 902 of FDASIA	• 21 CFR part 314, subpart H • 21 CFR part 601, subpart E • Section 506(c) of the FD&C Act, as amended by Section 901 of FDASIA	Prescription Drug User Fee Act of 1992 (PDUFA)

[a] Guidance for Industry: Expedited Programs for Serious Conditions – Drugs and Biologics (May 2014)

or therapies for the condition exist, offering significant improvements over current therapies.

The term *serious condition* simply refers to the severity of the disease and how it impacts the patient's quality of life, including persistent morbidity, the potential for deterioration if the condition is not remedied, and the potential for death.[4] When a serious condition is not adequately addressed by the current standard of care (or when treatment is nonexistent), then it qualifies as an *unmet medical need*. This can be demonstrated early in product development through nonclinical and pharmacologic data, a justification for the product's mechanism of action, or clinical evidence in later stages. In cases where *existing treatments* are available, the sponsor's proposed product and the FDA-approved product must share the same indication for use and demonstrate superiority in areas such as effectiveness, minimized side effects, or addressing emerging or unforeseen public health needs. Because the development landscape for a product evolves, these considerations should be integral to early discussions with the agency. Supporting clinical literature, expert recommendations, and guidance from regulatory agencies and scientific bodies further inform the agency's decision on whether the product meets these conceptual criteria.

The burst of innovation in biotechnology, along with increased focus on niche indications or specific patient subgroups, has prompted the agency to create new pathways to accommodate such development programs. These designations may qualify sponsors for some or all the benefits of the four expedited pathways, in addition to other incentives such as tax exemptions and market exclusivity. The following programs are discussed in this chapter:

- Orphan Drug Designation
- Regenerative Medicine Advanced Therapy Designation
- Rare Pediatric Disease Designation
- Qualified Infectious Disease Product Designation
- Breakthrough Device Designation

EXPEDITED PATHWAYS AND OTHER PRODUCT DESIGNATIONS IN DETAIL

The four key expedited programs are outlined below in the order that sponsors typically apply for them during the product development lifecycle. Generally, a regulatory strategy to expedite product approval begins with *Fast Track Designation*, followed by *Breakthrough Therapy Designation* if the sponsor can demonstrate that the product meets the necessary criteria. These two programs are instrumental in ensuring the FDA dedicates attention and resources to the product, fostering robust interactions to address and mitigate concerns during development. *Accelerated Approval* and *Priority Review* then expedite the review and approval of the marketing application, further streamlining the path to market.

Fast Track Designation

The FDA Modernization Act of 1997 established the Fast Track Designation program to streamline the drug development and approval process. It was expanded by the FDA Amendments Act of 2007 and the FDA Safety and Innovation Act of 2012 (FDASIA), which added provisions to further expedite the development and review of drugs that treat serious conditions and facilitate the early submission and review of data to the FDA. Fast Track Designation can be requested anytime during the development process, as early as the time of filing the Investigational New Drug application (IND).

The process of obtaining Fast Track Designation involves the following steps:

1. Assess Eligibility: Determine if the drug treats a serious condition and shows potential to address an unmet medical need.
2. Submit the Request: Prepare and submit a Fast Track Designation request to the FDA, accompanied by supporting data and relevant information.
3. FDA Review: The FDA evaluates the submission. If the drug is eligible, the FDA will notify the sponsor and provide guidance on the development plan and regulatory process.
4. Ongoing Collaboration: The sponsor continues working with the FDA to develop and evaluate the drug, benefiting from Fast Track perks such as more frequent meetings and priority review.
5. Expedited Review: The drug then undergoes the standard review process, with the goal of expediting approval for those that meet Fast Track criteria

Key features for Fast Track Designation are as follows:

- Expedited review: Fast Track Designation allows for the expedited review of drug development and regulatory applications, with the goal of bringing treatments to patients more quickly.
- More frequent communication: The FDA and the sponsor of the drug are encouraged to have more frequent meetings and exchanges of information during the development and review process.
- Priority review: Drugs that receive Fast Track Designation are eligible for Priority Review, which means the FDA will expedite the review of the application to get the drug to market faster.
- Rolling review: The FDA will review parts of the drug development application as they become available, instead of waiting for all the information to be submitted before beginning the review process. This typically starts around the time of the pre-NDA/BLA meeting with the agency.
- Accelerated approval: Fast Track Designation can lead to Accelerated Approval, which allows the FDA to approve drugs based on surrogate or intermediate endpoint data that is reasonably likely to predict clinical benefit.

- Flexible design: The Fast Track Designation program is designed to be flexible, allowing for changes in the development plan and regulatory strategy as needed to keep the drug development process on track.
- Continuous evaluation: The FDA will continuously evaluate the drug throughout its development to ensure that it remains eligible for Fast Track Designation and to identify any potential safety concerns.

Since its inception, the Fast Track Designation program has been implemented to help accelerate the development and approval of numerous drugs that treat serious conditions, including cancer, HIV/AIDS, and other life-threatening diseases. Over 200 Fast Track Designation requests have been granted since 2013.[5]

Breakthrough Therapy Designation

The FDA enacted the Breakthrough Therapy Designation to expedite the development and review of drugs that show significant improvement over available therapies for serious or life-threatening conditions. Patients with these conditions often face limited treatment options and poor prognoses. Similar to Fast Track, Breakthrough Therapy is designed to accelerate the development and review of drugs and biologics to treat serious conditions or to satisfy unmet medical needs. However, while Fast Track may rely on nonclinical activity and theoretical rationale, Breakthrough Therapy *requires preliminary clinical evidence* showing substantial improvement over available therapies in a specific indication to qualify for this designation.

Preliminary clinical evidence is early data derived from Phase 1 and Phase 2 human trials, where a product sufficiently and credibly demonstrates considerable improvement in safety and/or effectiveness over existing treatment, or over placebo or well-documented historical control in cases where there is no available treatment. This evidence can be supported by nonclinical data. Studies can also compare the investigational product plus current standard of care to the standard of care alone. These types of initial comparative studies early in development are encouraged by the Agency.[6] Generally, the FDA does not expect preliminary clinical evidence to be conclusive but expects that data should show a clear difference in advantage over existing treatment.

Clinically significant endpoints can refer to a number of definitions: measurement of health outcomes or treatment effects that are relevant to the specific patient population being studied and the medical intervention being evaluated; meaningful effects on established surrogate or intermediate endpoints that are predicted to provide clinical benefits or used to support traditional approval; or refer to significantly improved safety profiles, such as the reduction of serious adverse events, with evidence of equivalent or superior efficacy compared to existing treatments. What is deemed substantial improvement ultimately rests with the FDA, and this depends on the magnitude and duration of the effect on the endpoint and the observed effect on the treatment of the serious condition. Ultimately, these endpoints should be meaningful, reliable, reproducible, and based on well-established medical criteria, and their use in research helps to ensure that medical interventions are effective and provide real benefits to patients.

Breakthrough Therapy Designation status provides several benefits to the drug development process, including:

- Receiving priority review in the FDA's review process, which can result in faster approval times.
- More frequent, intensive interactions with the FDA to help ensure the fastest possible development and approval.
- Receiving scientific advice from the FDA to help ensure the best possible development path. Organizational commitment involves assigning "senior managers and experienced review and regulatory health project management staff in a proactive, collaborative, cross-disciplinary review." The FDA may assign a cross-disciplinary lead project manager to facilitate the review of the development program and interactions with the sponsor and to serve as a scientific liaison between members of the review team.

The request for Breakthrough Therapy Designation can be submitted as early as the filing of the IND, but preferably no later than the end-of-phase-2 meetings to utilize any of the key features of the designation. This designation is intended to develop the evidence needed to support approval, so clearly it does not make sense to request it later in the development process or after the submission of an original application. It is important to note that once the designation is granted, the FDA also has the power to rescind it should subsequent data show that it is no longer supportive of the designation's statutory standard of approval. The FDA will respond to Breakthrough Therapy Designation requests within 60 days of receipt of the request.

Accelerated Approval

The Accelerated Approval program was established by the FDA as part of the Prescription Drug User Fee Amendments (PDUFA) to the Federal Food, Drug, and Cosmetic Act in 1992, largely in response to the AIDS epidemic. The program was designed to provide quicker access to new drugs for serious and life-threatening conditions while still ensuring the drugs are safe and effective. The FDA periodically updates the program to reflect new scientific and regulatory developments, and to ensure it continues to provide a balanced approach to the development and approval of drugs for serious and life-threatening conditions. For example, the 21st Century Cures Act, enacted in 2016, included provisions to further enhance the Accelerated Approval program by streamlining the drug development process and strengthening the FDA's ability to monitor and regulate drugs approved under the program. Additionally, the FDA has issued guidance documents over the years to provide additional clarity and direction on various aspects of the Accelerated Approval program, such as the use of surrogate or intermediate endpoints and the conduct of post-approval studies.

Clinical studies for drugs approved under the Accelerated Approval program may be shorter or smaller in size compared to traditional clinical trials. However, they are still required to provide sufficient evidence to support the approval of the drug for its intended use. Thus, the use of high-quality empirical evidence such as surrogate or intermediate endpoints, which are markers that can predict clinical benefit, can help expedite the approval process. The FDA accepts as empirical evidence "epidemiological, pathophysiological, therapeutic, pharmacologic, or other evidence developed using biomarkers…or other scientific methods or tools."

Surrogate endpoints are biological markers expected to predict clinical benefit, though they are not direct measures of clinical benefit. They are used as intermediate measures of efficacy when direct measures of clinical benefit, such as overall survival or quality of life, are not feasible within the trial's timeframe. Examples of surrogate endpoints include biomarkers, laboratory tests, vital signs, disease progression measures and response rates. It is important to note that the use of surrogate endpoints must be carefully validated and considered in the context of the disease being treated and the stage of development of the drug.

Intermediate clinical endpoints are markers that predict clinical benefit but are not ultimate endpoints like improved survival or reduced disease symptoms. They provide an alternative for assessing drug efficacy when direct measures are not practical. Examples include changes in tumor size, disease severity, functional improvement, and response rates. The FDA recognizes that intermediate clinical endpoints can provide patients with earlier access to new treatments while providing necessary information needed for evaluating the drug's safety and efficacy.

It is important to note that the FDA will not grant Accelerated Approval to products that meet standards for traditional approval. The onus is on the sponsor to discuss the possibility of Accelerated Approval early in the regulatory development stage. The program is reserved for products with rigorous evidence to treat a serious condition and provide a meaningful advantage over existing treatment.

Sponsors must submit a New Drug Application (NDA) or a Biologics License Application (BLA) to the FDA, along with evidence to support the drug's eligibility for the program. The FDA reviews the evidence based on disease severity, the unmet medical need, and the quality and strength of the evidence. The application must meet the same statutory standards for safety and effectiveness as those granted traditional approval, and demonstrating safety and efficacy through adequate and well-controlled clinical trials under proposed conditions. Moreover, evidence of pharmacologic or biologic activity alone is insufficient; it must correlate with a clinical outcome. Sponsors must submit all promotional materials, including labeling and advertisements, during the pre-approval review period and at least 30 days before initial dissemination.

The sponsor must also conduct post-approval studies to confirm the clinical benefit of the drug and to ensure its ongoing safety and effectiveness. Typically, these confirmatory trials should proceed as soon as possible if it is clear that a surrogate or intermediate clinical endpoint can be used as a basis for approval, even prior to the submission of the marketing application. If the post-approval studies do not demonstrate the expected clinical benefit, the FDA may take action to restrict or remove the drug from the market.

Lastly, the sponsor must prepare for a rapid pace of development and commercial launch, including validation and scaling up its chemistry, manufacturing, and controls (CMC) as well as the development of packaging, labeling, promotional materials, and any other necessary companion products or diagnostics.

Priority Review Designation

The Priority Review program was enacted in the inaugural PDUFA in 1992 to expedite the development and approval of drugs that offer significant advances in treatment or provide a treatment where no adequate therapy exists for serious or life-threatening conditions. To avert the lengthy review process that was then typical of drug approval, the provisions in PDUFA provided specific benchmarks for the FDA to review and approve regulatory applications within a shorter timeframe. Essentially, a two-tiered paradigm was created, one for products that meet the criteria for Priority Review, and the other for everything else under the Standard Review process. The FDA set a target review time of six months for drugs that receive a Priority Review designation, compared to the standard review time of ten months for most other drugs. In 2015, after more than two decades of implementation, the FDA improved its metrics so that standard review and approval times for drug applications are now 12 months on average, while Priority Review applications have been efficiently reduced to eight months.[7] To be eligible for Priority Review, a drug must provide a treatment option for a disease or condition where there is a significant unmet medical need and must provide a meaningful advantage over available treatments; demonstrate the potential to provide a significant improvement in efficacy or safety over existing treatments for the target disease or condition; have the potential to offer a significant impact on the diagnosis, treatment, or prevention of a serious or life-threatening disease or condition; involve a new mechanism of action, new technology, or a new approach to treatment; and must demonstrate substantial evidence of effectiveness and safety.

Sponsors can request this designation at the time of NDA or BLA submission. The FDA informs the applicant of a Priority Review designation within 60 days of the receipt of the original BLA, NDA, or efficacy supplement.

Orphan Drug Designation

The Orphan Drug Designation program was established by the FDA in 1983 as part of the Orphan Drug Act to encourage the development of treatments for rare diseases and conditions, also known as "orphan diseases." It was signed into law to provide financial and regulatory incentives to encourage drug companies to develop treatments for these diseases following growing concern that few treatments were being developed because these conditions often lacked a financial return for the companies due to their small patient populations. Prior to this federal law, only ten industry-supported products for rare diseases had been approved and commercialized.

Today, over 600 orphan drug indications have been approved from over 450 distinct products.[8]

The ODA and the Rare Diseases Act of 2002 define rare diseases strictly according to prevalence, specifically serious conditions that affect fewer than 200,000 people in the United States. This is the key criterion to be eligible for orphan drug designation, which is given after a sponsor submits the necessary information required in 21 CFR 316. Additionally, the drug must not have been previously approved for the same use and must not have an active application for the same use pending with the FDA.

If the drug is granted Orphan Drug Designation, the company will receive certain benefits, including:

- Enhanced patent protection and marketing rights after approval, including seven years of market exclusivity, meaning no other drug can be approved for the same use during that time period.
- Waived or reduced user fees (i.e., not subject to PDUFA fees unless the application includes an indication for something other than the rare disease or condition for which the drug was designated).
- Tax credits for research and development costs, up to 50% for expenditures incurred during the clinical trial phases, for both approved and non-approved orphan products, allowing manufacturers to carry unused credits forward 20 years and back one year when a research-stage company may start earning income.
- Scientific and regulatory assistance from the FDA, including features of Fast Track Designation and Priority Review, for protocol assistance and to accelerate the development and approval process.
- Potential clinical research subsidies and funding from the Orphan Products Grants Program, non-profits (public-private partnerships), and private organizations.

Having an orphan designation, however, does not indicate that the product is safe or effective, nor does it alter the standard regulatory requirements and processes for obtaining marketing approval. The product must still show safety and effectiveness via adequate and well-controlled clinical studies, just like any other product.

Regenerative Medicine Advanced Therapy Designation

The Regenerative Medicine Advanced Therapy Designation, or RMAT, is a new designation designed for products in the regenerative medicine field that was born out of the 21st Century Cures Act. This federal law was enacted at the end of 2016 to accelerate the review of cell therapies, therapeutic tissue engineering products, human cell and tissue products, or any combination product using such therapies or products (except for those regulated solely under Section 361 of the Public Health Service Act (PHSA) and 21 CFR, Part 1271). In addition to this product classification, a product is also eligible for RMAT designation if the product is intended to treat, modify, reverse, or cure a serious or life-threatening disease or condition, and preliminary clinical evidence indicates that the drug has the potential to address unmet medical needs for such disease or condition.

The request for RMAT designation can be submitted concurrently with the filing of an IND application or as an amendment to an existing IND, though the FDA will not grant RMAT designation if the IND is on hold or is placed on hold during the designation review. Either way, a request for designation should be clearly stated on the cover letter of the IND application or amendment. The Office of Therapeutic Products (OTP) will notify the sponsor of the FDA's decision no later than 60 calendar days after receipt of the designation request. If OTP rejects the application, a written description of the rationale for such determination will be included in the notification to the sponsor. In general, the request for designation should be a concise package that includes

- A rationale for why the investigational product meets the definition of a regenerative medicine therapy.
- A description of the unmet medical need that the product has the potential to address and a justification to support why the intended disease or condition is considered serious.
- A summary of the risks and benefits associated with the therapies, if any, currently available for this condition.
- The preliminary clinical evidence, including the conditions for product administration, outcome assessment, and patient monitoring; a description of the patients and their outcomes, including the number of patients who have received the product; and the design, conduct, and analyses of any clinical investigations.

RMAT designation includes all the benefits and features of the Fast Track and Breakthrough Therapy expedited programs, including early interactions between the agency and sponsor to discuss potential surrogate or intermediate endpoints as well as clinical trial design, rolling review, intensive guidance on efficient product development as early as Phase 1, organizational commitment, and eligibility for Priority Review. However, unlike the Breakthrough Therapy designation, the RMAT designation does not require evidence to indicate that the product offers a substantial improvement over available therapies. Statutory standards for the demonstration of safety and effectiveness still need to be provided – the rigor of data collection; the nature and meaningfulness of the results; the number of subjects and sites; and the severity, rarity, or prevalence of the condition are all factors that feed into FDA's consideration.

As with the other expedited development programs, the RMAT designation can be rescinded if subsequent data or empirical evidence no longer qualifies it.

Rare Pediatric Disease Designation

The Rare Pediatric Disease Designation was established as part of FDASIA in 2012. Much like the Orphan Drug Designation program, it was established by the FDA to encourage the development of treatments for rare diseases that primarily affect children. To be eligible for Rare Pediatric Disease Designation, a drug must be intended to treat a disease or condition that primarily affects children and that affects fewer than 200,000 children in the US; must not have been previously approved for the same use; and must not have an active application for the same use pending with the FDA.

If the drug is granted Rare Pediatric Disease Designation, the company will receive certain benefits, including:

- Priority Review, which means that the FDA will expedite the review of the drug's application.
- Eligibility for a Pediatric Disease Priority Review Voucher, which can be redeemed to expedite the review of a subsequent drug application.
- Scientific and regulatory assistance from the FDA.

Qualified Infectious Disease Product Designation

Title VIII of FDASIA, *Generating Antibiotic Incentives Now* (GAIN), provides incentives for the development of antibacterial and antifungal drugs for human use intended to treat serious and life-threatening infections. Under GAIN, a drug may be designated as a *qualified infectious disease product* (QIDP) if it meets the criteria outlined in the statute. A drug that receives QIDP designation is eligible under the statute for Fast Track designation and Priority Review.

Breakthrough Device Designation

The expedited pathway for medical devices has been evolving. While approval of cutting-edge devices had come under the Expedited Access Pathway (EAP), Innovation Pathway, and Priority Review programs, the 21st Century Cures Act, via Section 3051, enacted by Congress in 2016, mandated the new Breakthrough Devices Designation to gradually supersede and fold the features of the previous expedited pathways for devices into one designation. Note that products that have previously been granted EAP designation will automatically be granted Breakthrough Devices designation, and no separate action is necessary to obtain it.

Breakthrough Device is a voluntary program for certain medical devices that are intended to *help patients have more timely access* to technologies that provide more effective treatment or diagnosis of life-threatening or irreversibly debilitating diseases or conditions. That is the primary qualification for the program. The device must also meet at least one of the following criteria:

- Represent breakthrough technologies.
- No approved or cleared alternatives exist.
- Offer clinically meaningful advantages over existing approved or cleared alternatives including the potential, when compared to existing approved

alternatives, to reduce or eliminate the need for hospitalization, improve patient quality of life, and facilitate patients' ability to manage their care.
- Device availability is in the best interest of patients

Under the Breakthrough Device program, the FDA works with sponsors to make the development and regulatory processes more efficient and cost-effective, expediting development, assessment, and review, while preserving the statutory standards required for premarket approval (PMA), clearance of a premarket notification (510(k)), and marketing authorization via the De Novo classification process to ensure safety and effectiveness. Components of the program include features from the Fast Track and Priority Review programs for drugs and biologics. The extent to which the FDA provides these features depends on the availability of resources. As in other designations and expedited pathways, the FDA strongly recommends that sponsors discuss the possibility of applying for this program early in the development of the device, particularly before starting an IDE pivotal study, so that there is agreement on clinical endpoints used and data collection methodologies appropriate for marketing approval.

The request for Breakthrough Device designation should be a standalone submission (i.e., no other requests are included with the request) and submitted as a Pre-Submission package to the appropriate review division. For combination medical devices, that could mean CBER instead of CDRH. Refer to FDA's guidance *Requests for Feedback on Medical Device Submissions: The Pre-Submission Program and Meetings with Food and Drug Administration Staff.*

CONCLUSION

These expedited pathways and product designations have evolved to meet the dynamic needs of patients, medical communities, and the FDA. Over the years, they have been instrumental in accelerating the development and review of life-saving treatments for serious conditions. As medical science advances, the FDA continuously refines and enhances these programs, integrating emerging technologies and innovative processes to further streamline drug development and approval. These ongoing efforts underscore the FDA's commitment to ensuring timely access to safe and effective therapies while adapting to the ever-changing landscape of healthcare.

RELEVANT REFERENCES

1. 21st Century Cures Act (Public Law 114–255) (December 2016) (114th Congress)
2. Breakthrough Devices Program: Guidance for Industry and Food and Drug Administration Staff (September 2023)
3. Expedited Access for Premarket Approval and De Novo Medical Devices Intended for Unmet Medical Need for Life-Threatening or Irreversibly Debilitating Diseases or Conditions (April 2015)
4. Expedited Programs for Regenerative Medicine Therapies for Serious Conditions (Draft) (November 2017)

5. Expedited Programs for Serious Conditions – Drugs and Biologics (May 2014)
6. Factors to Consider Regarding Benefit-Risk in Medical Device Product Availability, Compliance, and Enforcement Decisions (December 2016)
7. The Least Burdensome Provisions: Concept and Principles (Draft) (December 2017)
8. Qualified Infectious Disease Product Designation Questions and Answers (Draft) (January 2018)
9. Requests for Feedback on Medical Device Submissions: The Pre-Submission Program and Meetings with Food and Drug Administration Staff (September 2017)
10. Title 21, Code of Federal Regulations
    - **3 and 4: Combination Products**
    - **11: Electronic Records**
    - **312: Investigational New Drug Applications (IND)**
    - **314: Applications for FDA Approval to Market a New Drug (NDA)**
    - **316: Orphan Drugs**
    - **320: Bioavailability and Bioequivalence Requirements**
    - **601: Biologics (BLA)**
    - **812: Investigational Device Exemptions (IDE)**
    - **814: Premarket Approval (PMA)**

NOTES

1. "CDER Approval Times for Priority and Standard NDAs and BLAs Calendar Years 1993–2015" https://www.fda.gov/downloads/Drugs/DevelopmentApprovalProcess/HowDrugsareDevelopedandApproved/DrugandBiologicApprovalReports/NDAandBLAApprovalReports/UCM491533.pdf
2. Guidance for Industry: Expedited Programs for Serious Conditions – Drugs and Biologics (May 2014)
3. Guidance for Industry: Expedited Programs for Serious Conditions – Drugs and Biologics (May 2014)
4. 21 CFR 312.300(b)(1)
5. FDA-TRACK: Agency-wide Program Performance https://www.fda.gov/AboutFDA/Transparency/track/default.htm
6. Sponsors should consult the FDA's ICH guidance for industry E10 Choice of Control Group and Related Issues in Clinical Trials.
7. "CDER Approval Times for Priority and Standard NDAs and BLAs Calendar Years 1993–2015" https://www.fda.gov/downloads/Drugs/DevelopmentApprovalProcess/HowDrugsareDevelopedandApproved/DrugandBiologicApprovalReports/NDAandBLAApprovalReports/UCM491533.pdf
8. Insights into Rare Disease Drug Approval: Trends and Recent Developments (October 2017), https://www.fda.gov/downloads/forindustry/developingproductsforrarediseasesconditions/ucm581335.pdf

9 Investigator-Initiated Sponsored Research

Harvey M. Arbit and Wrenda Teeple

INTRODUCTION

Investigator-initiated and sponsored research (IISR) is also referred to as investigator-initiated trials (IIT), investigator-initiated studies (IIS), and investigator-initiated research (IIR). Investigator-initiated clinical research has long been a cornerstone of medical science and clinical innovation. Researchers may conduct clinical trials for myriad reasons, ranging from strictly academic purposes, such as publication for tenure, to gaining an understanding of certain aspects of the effect of a drug product to improve patient care, to furthering the development of a new drug substance towards the eventual submission of a New Drug Application (NDA), and obtaining the United States Food and Drug Administration (FDA)'s approval to market. Regardless of the underlying motivation of an investigator for undertaking clinical research, there are regulations that govern the entire process. Early in the translational research process, the most important regulations are those that involve investigational new drugs and devices.

Within the regulations governing both investigational new drugs and devices, there are sections specific to the obligations of investigators, such as the principal investigator (PI), and those that are specific to the obligations of the application sponsor, such as the holder of the Investigational New Drug Application (IND) or the Investigational Device Exemption (IDE). Some experienced researchers have served as PIs under commercial INDs. Under such an arrangement, the researchers do not have to concern themselves with the obligations of the IND or IDE sponsor; that is, the drug or device company (i.e., sponsor) that holds the IND or IDE. In the case in which a clinical researcher serves both the role of the investigator and sponsor – that is, as a sponsor-investigator (SI) – it is incumbent upon that researcher to be familiar with and adhere to the regulations that govern both the conduct of the investigator and the conduct of the sponsor.

The majority of biomedical and clinical researchers, however, have little experience with these regulations, as opportunities for formal training are few and far between. In fact, they may not be fully aware that they have assumed the role of both sponsor and investigator for a particular study, and therefore may not fully comprehend the clinical, regulatory, financial, contractual, and ethical risks for which they assume responsibility as the SI of a clinical trial. Additionally, investigators who

DOI: 10.1201/9781315117911-9

choose to take on the role of SI may not be fully aware of the support and research infrastructure necessary to successfully conduct the research under an IND or IDE. In this chapter, we will explore the importance of investigator-initiated sponsored research (IISR), the role of the SI in clinical trials, regulatory oversight of trials conducted by SIs, and key considerations for investigators choosing to take on the dual role of SI.

WHY CONDUCT INVESTIGATOR-INITIATED RESEARCH?

When discussing IISR, the investigator is usually a researcher at an academic institution, such as a university, academic medical center, or teaching hospital. These researchers are uniquely positioned to understand pressing clinical questions and to conduct the translational research and clinical trials needed to bring new discoveries and therapies to the clinic. Furthermore, these individuals often have experience working with patients and have access to potential trial participants. In these settings, the initiation of clinical research may be the result of numerous motivating factors that may simultaneously serve the needs of the patient and the needs of the researcher's career.

Among the more altruistic motivations are the desires to pursue one's scientific curiosity and to improve patient care. This may include finding more efficient ways to facilitate the administrative aspects and improve healthcare delivery itself, such as implementing improved electronic medical record systems. Researchers may be eager to discover a new molecule, dosage form, or delivery system to improve patient outcomes, seek to repurpose an existing drug to treat a different disease or population, or make drugs easier and more convenient for patients to take to increase adherence.

The reality of the academic research environment, however, means that these beneficent desires are often tempered by the need to achieve specific milestones throughout one's career, such as promotion and tenure, and the need to secure funding to sustain one's research program. Academic institutions require faculty to conduct research and publish their results to qualify for promotion and tenure. In order to conduct this research, it is necessary for researchers to successfully apply for and be granted funding, which often comes from federal funding sources, such as the National Institutes of Health (NIH). The catch 22 for researchers early in their careers is that many funding mechanisms require the researcher to have a track record of both publications and prior funding; this places the early-career researcher at a disadvantage when it comes to being awarded a grant, as they have not yet had the time or resources to develop the required track record. In these instances, it is often advantageous for the researcher to pursue funding from a commercial entity. As a researcher's career matures, he or she successfully competes for and is awarded funding, leading to recognition as a key opinion leader (KOL). For those researchers, winning additional grants becomes easier, as does being invited to participate in a commercially sponsored IND or IDE.

When a researcher plans to undertake IISR, there are several considerations that are necessary to examine prior to beginning any studies or trials. First, the researcher needs to determine whether the primary objective of the work is to continue to study the product for academic reasons or to bring the product to market. Second, the researcher needs to clearly identify their role in the clinical research project; that is, are they serving as an investigator, sponsor, or both? Having a clear understanding of the objectives and sponsorship of the project early on is critical in determining research strategies, regulatory approaches, designing the appropriate studies, optimizing the use of resources, and identifying the strongest potential funding sources.

SPONSOR, INVESTIGATOR, OR SPONSOR-INVESTIGATOR?

As discussed, there are two determinations that must be made by a researcher prior to undertaking clinical research: what the ultimate goal is for the product under investigation and which parties are serving as the sponsors and/or investigators. The former is much more easily determined than the latter. In this next section, we aim to provide some clarity regarding how the terms "sponsor," "investigator," and "sponsor-investigator" are defined, differences in funding paradigms for an "investigator" versus a "sponsor-investigator," and the regulatory and monitoring responsibilities of each role.

DEFINING "SPONSORS" AND "INVESTIGATORS"

According to the relevant sections of the Code of Federal Regulations (CFR), a clinical investigation is "any experiment in which a drug is administered or dispensed to, or used involving, one or more human subjects" (OFR 2019l). Experiment is defined as "any use of a drug except for the use of a marketed drug in the course of medical practice" (OFR 2019l). According to the FDA/HCFA Interagency Agreement, an experimental (Category A) device is an

> innovative device believed to be in Class III for which 'absolute risk' of the device type has not been established, i.e., initial questions of safety and effectiveness have not been resolved, and FDA is unsure whether the device type can be safe and effective. A nonexperimental/investigational (Category B) device refers to a device believed to be in Class I or II, or a device believed to be in Class III for which the incremental risk is the primary risk in question. (USFDA, CDRH, CBER 2017)

An investigational device is defined as "a device that is the object of an investigation" (OFR 2019r). An investigational new drug means a new drug or biological drug that is used in a clinical investigation (OFR 2019l). When discussing roles of key individuals involved in the conduct of clinical research through a regulatory lens, it can be difficult to understand who serves as the "investigator," who serves as "sponsor," and whether those individuals are, in fact, one and the same (that is, a "sponsor-investigator").

An investigator is "an individual who actually conducts a clinical investigation" (OFR 2019r). From a regulatory perspective, a sponsor, on the other hand, is an individual or company "who takes responsibility for and initiates a clinical investigation" but "does not actually conduct the investigation unless the sponsor is a sponsor-investigator" (OFR 2019l, 2019r). A sponsor may also be viewed from a financial perspective, adding to the above definition that they "provide funding and/or product," but do not serve as the regulatory sponsor (OFR 2019e). In the case of a commercially-funded IND/IDE, the "financial and regulatory sponsors are usually the same"; in an "investigator-initiated IND/IDE," however, the investigator usually also serves as the regulatory sponsor, though not necessarily the financial sponsor (OFR 2019e). If an investigator both "initiates and conducts an investigation," then they are referred to as the SI; note that this can refer only to an individual person, not a group of researchers or a company (OFR 2019l). In some cases, it may be challenging to determine whether the PI is serving as an investigator or as a sponsor-investigator. Answering the questions listed in Box 2.1 can provide insight into whether an investigator is also the sponsor of a clinical investigation. In general, a researcher participating in a study sponsored by a commercial entity would be a PI at their institution and would not dually serve as a regulatory sponsor. In such a case, that responsibility rests with the commercial sponsor that holds the IND or IDE.

BOX 2.1 Investigator or Sponsor-Investigator?

The following questions can be used to help determine whether an investigator is also the sponsor of an IND/IDE study (i.e., the sponsor-investigator):

1. Was the study protocol developed by the researcher (with or without input from the product's company)?
2. Did the researcher request financial and/or product support from a company?
3. Will the researcher be the study investigator?
4. Will the investigator also hold the IND/IDE?
5. Was an assessment conducted that determined that 21 CFR 312 or 21 CFR 812 applies to the study?
6. Has the company expressed that it has no intention of being the regulatory sponsor?
7. If the answer to most of these questions is "yes," then the investigator is also likely the regulatory sponsor of the study.

There are exceptions, however, including select scenarios in which a study may be initiated by a company, but the investigator is viewed by regulatory agencies as serving as the sponsor-investigator. Three examples are provided in Table 9.1. Scenario 1 is *truly investigator initiated*, in that the study idea and protocol originate with the researcher, the study aims to answer an academically-oriented question, and the company providing financial support and/or the product for testing has no interest in the outcome; thus, the researcher is the sponsor-investigator. In Scenario 2, the study idea and protocol originate with the investigator (with potential input from the company) and the study aims to answer an intellectual question. This is *investigator initiated, with a twist*, in that the company has an interest in the study outcome and wants access to the results. In this context, the researcher again serves as the sponsor-investigator. When a researcher finds himself/herself in a situation like Scenario

TABLE 9.1
Scenarios in Which the Researcher Serves as Sponsor-Investigator

Question	Scenario 1	Scenario 2	Scenario 3
Whose idea was the clinical investigation?	Investigator	Investigator	Company
Who developed the protocol?	Investigator	Investigator, with or without company input	Company, with or without investigator input
Does the investigator hold the IND/IDE?	Investigator, if necessary	Investigator, if necessary	Investigator, if necessary
What is the purpose of the investigation?	Answer intellectual question	Answer intellectual question	Collect data
Who provides the financial support/product?	Company or other	Company or other	Company
Does the company have an interest in the outcome?	No	Maybe; wants access to results	Yes; wants access to results
Is the investigator the sponsor?	Yes *(investigator initiated/investigator sponsored)*	Yes *(mostly investigator initiated/investigator sponsored)*	Yes *(company initiated/investigator sponsored)*

2, it is important to ensure that they communicate with the company in advance of the start of the investigation to ensure that ownership and use of data are clearly established. Scenario 3 details a more complex arrangement, in which the company appears to be the sponsor of the study, as the study idea and protocol originate with the company, and the company has an interest in the results.

The investigator and their institution, however, may be financially responsible for monitoring the investigation (sponsor obligation), thus making this study *investigator-initiated in disguise*. In short, transparency is of the utmost importance, regardless of the complexity of the study arrangements between the researcher and the company.

Funding Investigator-Initiated Research

The source of funding for IISR depends greatly on who developed the product under study and can, in part, dictate who holds the role of financial sponsor. If the company with which the investigator is partnering developed the product, then the monetary and/or product support usually comes from that particular company. This includes research conducted on an existing marketed product, whether used as supplied or modified by the researcher. If the researcher has developed a novel product not commercially available or manufactured by an outside company, however, the funding likely will originate from a foundation or government entity.

During the process of requesting and negotiating study support with investigators, the companies examine several factors, as they maintain specific policies, procedures, and budgets for supporting investigator-initiated research. One key factor in dividing resources efficiently and equitably among investigators is the determination of fair market value (FMV) detailed in Box 2.2. For example, the cost to conduct a clinical study in New York City might be much greater than if the same study were to be conducted in Butte, Montana. These regional differences play an important role in the distribution of investigator-initiated research dollars.

BOX 2.2 Assessing Fair Market Value

Fair Market Value, or FMV, is defined as the price at which buyers and sellers are "willing to do business," taking into account "reasonable knowledge of pertinent facts" and "not acting under any compulsion."

When assessing FMV for a drug or device, the company must develop a structure to support spending decisions, implement a consistent and repeatable process for all investigator-initiated and sponsored research, and evaluate scientific merit. Ultimately, the product must align with the company's research and development strategy.

When conducting and presenting research, the investigator must disclose all financial support of their research and must not exceed local FMV assessments. The funding provided by a company must not be used for expenses that are not associated with the research or not included in the FMV assessments.

Regulatory Oversight and Monitoring

The greatest factor in determining sponsorship in the eyes of regulatory bodies, however, is identifying the appropriate regulatory sponsor. For example, in the scenarios detailed in Table 9.1, the company does not necessarily serve as the regulatory sponsor of the IND or IDE, as they are not assuming responsibility for monitoring, notifying the FDA of serious adverse events (SAEs), and regulatory compliance; rather, they are simply the study support resource or financial sponsor. In such cases, the researcher is an SI. On the other hand, in a study in which the company explicitly assumes regulatory sponsorship on the IND/IDE application, the participating researcher would serve only as PI at their institution. In short, whoever holds the sponsorship on the IND/IDE assumes responsibility for initiating and monitoring the conduct of the clinical investigation. The FDA's role is to protect the health, safety, and welfare of the patients and research participants. Complying with FDA regulations during the conduct of clinical trials is of utmost importance. Failure to comply with the obligations of an SI under an approved IND/IDE can lead to a Warning Letter. Failing to obtain an IND or IDE for a clinical investigation can also lead to a Warning Letter. In clinical investigations, consequences for non-compliance can be significant and, in some situations, career-ending for the PI or SI. All official Warning Letters sent by the FDA are publicly available on the FDA website and often include the name of the investigator, sponsor, and/or company as well as other detailed information related to the failure of the sponsor and/or investigator to comply with regulatory obligations and the protection of human research subjects.

RESPONSIBILITIES OF THE SPONSOR-INVESTIGATOR

Clinical researchers who become sponsor-investigators are often unfamiliar with FDA regulations. Those with previous experience serving as PIs under commercial INDs may have some familiarity with the obligations of an IND principal investigator, but not necessarily those of an IND sponsor. While an overview of the responsibilities a researcher assumes as an SI is detailed here, it is highly recommended that investigators considering serving as an SI consult with or include a regulatory professional as a member of the research team. This arrangement allows the investigator to focus on the research with which he/she is familiar, while the regulatory professional can ensure that the SI is fulfilling all the required FDA obligations.

Regulatory Expectations and Obligations

When a researcher initiates a study for which they serve as a sponsor, several key forms are signed during the IND or IDE application to the FDA. Each form commits the investigator to key sponsor responsibilities dealing with oversight of study conduct, adherence to relevant regulations, and monitoring, and they are collectively excellent windows into the obligations of an SI. Researchers initiating INDs as SIs are required to sign two forms, Forms FDA 1571 and 1572 (Table 9.2). It is of the utmost importance that researchers initiating a study for which they will serve as

TABLE 9.2
Responsibilities of Sponsors and Investigators of IND and IDE Clinical Investigations, as Outlined by the FDA

	IND (21 CFR 312, Subpart D)		IDE (21 CFR 812, Subparts C, D, E, G)	
Responsibilities	**Form 1571 (Sponsor)**	**Form 1572 (Investigator)**	**Sponsor**	**Investigator**
General Responsibilities	312.50	312.60	812.40	812.100
Control of Investigational Drug	–	312.61	–	–
Recordkeeping and Record Retention	312.57	312.62	812.140	812.140
Investigator Reports	–	312.64	–	–
Assurance of IRB Review/Approval	–	312.66	812.62	–
Inspection of Records and Reports	312.58	312.68	812.145	812.145
Handling of Controlled Substances	–	312.69	–	–
Transfer Obligations to CRO	312.52	–	–	–
Selecting the Investigators and Monitors	312.53	–	812.43	–
Emergency Research	312.54	–	812.47	–
Informing Investigators	312.55	–	812.45	–
Review of Ongoing Investigations	312.56	–	812.46	–
Disposition of Unused Investigational Drugs or Devices	312.59	–	812.46	–
Specific Responsibilities of Investigators	–	312.61–312.69	–	812.110
Disqualification of a Clinical Investigator	–	312.70	–	812.119
UAE Safety Reports	312.32	312.32	812.150(b)(1)	812.150(a)(1)
Annual Progress Reports	312.33	312.64	812.150(b)(5)	812.150(a)(3)
FDA and IRB Approval	312.40	312.66	812.42	–
Monitoring Investigations	312.56	–	812.46	–
Significant Risk Device Determinations	–	–	812.66	–
Labeling	312.6	–	812.5	–
Prohibitions Against Promotion	312.7	–	812.7	–

both sponsor and investigator read and understand these forms in their entirety before signing. Both forms, FDA 1571 and FDA 1572, include the statement "WARNING: A willfully false statement is a criminal offense (U.S.C. Title 18, Sec. 1001)," underscoring the importance of adherence to these regulations in their entirety.

Form FDA 1571 applies to sponsors of IND investigations, and outlines the obligations associated with sponsorship (Table 9.2). Signing Form FDA 1571 commits the sponsor or sponsor-investigator to wait 30 days before beginning the study for a mandatory safety review of an initial IND application by FDA; to wait to continue the study if placed on clinical hold; to obtain Institutional Review Board (IRB) approval of the study and informed consent document; and to conduct the study in accordance with all applicable regulatory requirements. Included in this is ensuring that monitoring is done appropriately, that the protocol and IND are current and in compliance, and that the sponsor informs the FDA of SAEs (Box 2.3). In company-initiated and sponsored studies, this form represents a contractual agreement made between the company and the federal government; in investigator-initiated studies, Form FDA 1571 is signed by the researcher undertaking the study and represents a contractual agreement between the researcher and the federal government.

BOX 2.3 Reporting Serious Adverse Events

Serious Adverse Events (SAEs) are defined as "any untoward medical occurrence associated with the use of a drug in humans" which results in "death, a life-threatening adverse event, inpatient hospitalization or prolongation of existing hospitalization, a persistent or significant incapacity or substantial disruption of the ability to conduct normal life functions, or a congenital anomaly/birth defect" (USFDA 2020).

In *investigator-initiated research*, the investigator is responsible for submitting the SAE information to the IRB. As they also serve as the IND/IDE sponsor, they must also submit the SAE information to the FDA.

In *commercial research*, the investigator is responsible for submitting the SAE information to the IRB. As the investigator is not the sponsor, they must also submit the SAE information to the commercial IND/IDE sponsor. The commercial sponsor then submits it to the FDA.

The second form, Form FDA 1572, details the requirements of an IND study investigator, including conducting the study per protocol, personally supervising or conducting the investigation, informing subjects of the investigational status of the test article, reporting adverse events to the sponsor, reading and understanding the Investigator's Brochure, and informing all support personnel of investigation requirements (Table 9.2). It also commits the investigator to being the point-of-contact with and assuming responsibility for IRB interactions on behalf of the study, namely, to maintain adequate records and make them available for inspection, assure that the IRB is in compliance, assume responsibility for initial and continuing IRB review, only make changes to the research protocol with the approval of the IRB, and comply with obligations of clinical investigators.

The expectations of IDE sponsors and investigators share many similarities with those involved in IND investigations; however, there are key differences, due in part to differences between drug and device studies. Similar to sponsors of drug studies, sponsors of device-focused investigations are responsible for ensuring monitoring, keeping the protocol and IDE current and compliant, and informing the FDA and IRB of significant new information (Table 9.2). Sponsors of IDEs are also responsible for device risk determination, seeking and maintaining IRB approval, labeling, filing annual progress reports, and filing unexpected adverse events (UAEs). Investigators involved in IDE studies share some of the same general responsibilities, as well as adhere to specific ones; these include awaiting approval, compliance, supervising device use, financial disclosure, and disposing of devices (Table 9.2). Researchers serving as sponsor-investigators for an IDE study must comply with both sets of regulations, i.e., sponsor and investigator obligations.

Device Risk Determination

The sponsor of an IDE study holds the responsibility to determine the risk level of the device under study. As the sponsor-investigator of an investigational device study, the researcher is responsible for making the initial risk determination and presenting the risk assessment to the IRB. A device is considered to be a significant risk (SR) device if intended as an implant, supports or sustains life, plays a substantial role in diagnosing, curing, mitigating, or treating disease, or whose use has the potential to cause serious harm from either the implantation procedure or the device itself. The initial study of SR devices requires approval from both the IRB and the FDA via the IDE mechanism. A non-significant risk (NSR) device is one that does not meet the requirements of an SR device; NSR should not be confused with "minimal risk," which is an IRB term meaning that the probability and magnitude of harm or discomfort anticipated in the research are not greater in and of themselves than those ordinarily encountered in daily life or during the performance of routine physical or psychological examinations or tests (OFR 2019d). In studying a device deemed NSR, the FDA requires that abbreviated IDE sponsor regulations be followed, and that the IRB approves the study and acts as the "surrogate overseer" (USFDA 2006). Abbreviated IDE sponsor requirements include ensuring that the study device is labeled appropriately, obtaining and maintaining IRB approval, obtaining and documenting informed consent, appropriately monitoring the investigation, maintaining records and reports, and complying with the prohibitions against promotion, all per the appropriate sections of the CFR.

In determining the risk of a device, the SI must account for several study factors. First, the risk determination needs to be based on the proposed use of a device in the investigation, and not simply the device alone. This means that risks associated with any procedures involved in the implantation, use, or removal of a device must be considered. Second, the SI must consider that SR studies present potential for serious risk to the health, safety, and/or welfare of a subject, and such risks must be thoroughly and explicitly detailed. Third, the IRB must review the SI's SR or NSR determination for every investigational medical device study and reassess risk for every study modification review. Finally, the FDA is the final arbiter as to whether a device under study is SR or NSR. For more information, please see the FDA guidance, Information Sheet Guidance for IRBs, Clinical Investigators, and Sponsors: Significant Risk and Nonsignificant Risk Medical Device Studies (USFDA 2006).

As the IRB plays a critical role in the determination of the risk of an investigational device, there are several expectations of the board. The IRB should have standard operating procedures that explain how the IRB makes an SR and NSR determination and explain how the final decision should be documented. The IRB should make the SR or NSR determination about a study through a review of the relevant information at an officially convened meeting. The board has the right to agree or disagree with the SI's initial assessment. If the IRB determines that the study is NSR, it can approve the study, and the study may begin without submission of an IDE application to the FDA, as an IDE is considered to be in effect. If the IRB disagrees with an NSR assessment and deems that the device and study are SR, the IRB must inform the clinical investigator, sponsor, and/or SI. If the IRB determines that the study is SR, it may approve the study as an SR device study, but the study may not begin until the FDA approves the sponsor's IDE application. It is under the purview of the IRB to ask the sponsor or SI for proof that the SR study's IDE application has received FDA approval, usually in the form of a copy of the approval or conditional approval from the FDA.

Clinical Trial Registration

One of the regulatory requirements set forth by the FDA for sponsors is to register all clinical trials at ClinicalTrials.gov (OFR 2019c). ClinicalTrials.gov is a database maintained by the National Library of Medicine at the U.S. National Institutes of Health, which makes information regarding publicly and privately supported clinical studies of human participants conducted globally available to the public. It was created by the FDA Modernization Act (FDAMA) of 1997, Section 113, which required that NIH create, maintain, and operate a publicly available resource to track drug efficacy studies that were conducted under Investigational New Drug (IND) applications. The purpose of creating a database was to improve the public's knowledge and awareness of clinical drug trials for individuals with serious or life-threatening conditions in which experimental treatments might be helpful. In 2007, Section 801 of the FDA Amendments Act (FDAAA) required the listing of drug and device clinical trials in ClinicalTrials.gov. Certification of Compliance with Requirements of ClinicalTrials.gov Data Bank, or Form FDA 3674, was developed and became a requirement for most IND and IDE clinical studies (OFR 2019k). The instructions for filling out Form FDA 3674 Certification of Compliance describe in detail how to complete Form FDA 3674 (USFDA 2017).

It is the study sponsor's responsibility to register the study on ClinicalTrials.gov, notify the FDA using Form FDA 3674, maintain the record, and provide a summary of the results upon completion of the study. For investigators who serve as sponsors of IISR studies, it is their obligation to register their investigation and maintain the study entry in ClinicalTrials.gov. Applicable clinical trials are those involving drugs, biologics, and devices overseen by the FDA. These trials must be registered, particularly if their endpoints are related to health outcomes; this includes IND and IDE studies and studies with one or more sites in the United States. Failure to register and submit required clinical trial information for applicable trials is considered a violation of the law as part of 21 CFR Part 331 Prohibited Acts. Studies that use non-serious and non-life-threatening drugs to examine disease processes or employ a behavioral intervention, feasibility trials of prototype devices (i.e., do not have health outcomes as primary endpoints), and expanded access single patient INDs are generally exempt from this requirement. Once a clinical trial is registered at ClinicalTrials.gov, it is assigned a National Clinical Trial (NCT) number. Clinical trials with an NCT number can be linked to a journal article with a PubMed identification number (PMID). In 2004, the International Committee of Medical Journal Editors (ICMJE) issued a policy requiring the listing of clinical trials in a public database. Without such a listing, the research manuscripts would not be considered for publication, regardless of their importance to the scientific and medical community.

Clinical Trial Monitoring

Monitoring clinical investigations of new drugs and devices falls under the purview of the sponsor or SI of an IND/IDE study. The term "monitoring" in the context of a clinical trial refers to "a quality control tool for determining whether study activities are being carried out as planned, so that deficiencies can be identified and corrected" (USFDA, CDER, CBER, CDRH 2013). Such activities ensure the safety of human subjects enrolled in an IND or IDE study and that the data produced from the study are as high-quality and robust as possible.

There are several approaches to monitoring a clinical investigation. Regardless of which approach is selected, however, two things must be determined *a priori* to ensure that monitoring is done consistently and thoroughly: the establishment and maintenance of standard operating procedures (SOPs), and identification of the person(s) actively providing oversight and monitoring study conduct. SOPs ensure that monitoring is done in a structured fashion, such that the monitoring process would be consistent if multiple trained study monitors were employed. Additionally, the FDA has the right to inspect SIs to determine if and how study monitoring was done and may request to see monitoring SOPs. If any changes are made to the monitoring SOPs, they must be documented in a controlled manner such that all changes are traceable over the course of the study.

To reiterate, monitoring is a sponsor's obligation; in IISR studies, the SI is held responsible for all study conduct. Therefore, it is critical to select reliable individuals to oversee study monitoring and to ensure that these representatives are appropriately trained. This raises the question as to who can monitor a study. An SI may monitor the study if they are qualified by training and experience; however, this may present a conflict of interest and may be disallowed by the FDA. The use of research staff presents a similar problem. Therefore, it is recommended that an SI either use their institution's monitoring service or that they contract with a Contract Research Organization (CRO). For INDs, the transfer of Clinical Trial Monitoring responsibilities from an SI to a CRO should be done in writing. Any responsibilities not transferred in writing remain with the sponsor. By using the monitoring services provided by institutional services or an autonomous group outside of the SI's institution, the conflict of interest concerns and the appearance of impropriety are adequately addressed. Self-monitoring by the SI is usually considered unacceptable. Should the SI be inspected by the FDA, it will be the FDA investigator's discretion if the self-monitoring activities are acceptable. While the monitoring activities for investigational devices may be conducted by a CRO, the transfer of clinical trial monitoring responsibilities to a CRO is not allowed for investigational devices, and the sponsor will ultimately be held responsible for all required monitoring.

Sourcing Investigational Drugs

When identifying the source for a therapeutic agent for an IND, there are several factors that the sponsor must consider. First, it is important to understand that the FDA views any drug not produced using current Good Manufacturing Practices (GMPs) as adulterated, and therefore not admissible in a clinical investigation. In Phase I trials, this requirement is statutory and must be outlined in well-defined, written procedures. Manufacturing drugs under GMPs means that the equipment used in the manufacturing process is adequately controlled, as is the environment in which the fabrication process takes place. Furthermore, data must be accurately, consistently, and contemporaneously recorded throughout the manufacturing process, including any testing that may take place. In Phase II and III trials, the use of GMPs in production is a regulatory requirement (OFR 2019h, 2019i; USFDA 1991, 2008).

If the drug under study is already lawfully on the market, it may be exempt from IND regulations if it meets the following additional requirements (OFR 2019j). First, the study must not be conducted with the intention of supporting a submission to the FDA for a new indication (OFR 2019b, 2019f). Second, the study must not be used in a submission to the FDA to support an advertising change (OFR 2019b, 2019f). Third, the study must not involve a change in the route of administration of the drug, its dosage level, use in a patient population, or other manner that may increase the risk of the product (or decrease the acceptability of the risks) as currently approved by the FDA (OFR 2019b, 2019f). Fourth, the study must be conducted in compliance with the institution's IRB and have appropriate informed consent procedures (OFR 2019b, 2019f). Finally, to be exempt from IND regulations, the study must not promote the drug or represent it to be safe and effective. The SI may not charge the patient or the patient's insurance for the investigational drug unless the SI has requested and received prior written authorization from the FDA to do so (OFR 2019o). If these criteria are met, then the study drug is not subject to IND regulations. Key guidance documents detailing whether research studies require an IND are listed in Box 2.4, along with several other helpful guidance documents.

When conducting clinical investigations in conjunction with a company supplying the investigational product, it is critical that the sponsor-investigator understands from what supply of drug the study product will come. If the drug is already FDA-approved, the study product may come from a commercial supply but may have an auxiliary label indicating that it is for investigational use. Alternatively, an approved drug may be provided from a clinical supply, which could differ from the commercial supply in terms of where it was originally manufactured, GMP standards adhered to, testing, and material grade. If drawn from a clinical supply, the study drug would be labeled as being for investigational use; this is what is most frequently used in clinical investigations.

BOX 2.4 Useful Guidance Documents

The following FDA Guidance Documents may be of use in determining whether a sponsor-investigator's proposed study requires an IND or IDE application and approval. All can be found at www.fda.gov:

For IND applications:

- Investigational New Drug Applications: Determining Whether Human Research Studies Can Be Conducted Without an IND (UCM229175)
- IND Exemptions for Studies of Lawfully Marketed Drug or Biological Products for the Treatment of Cancer (UCM071717)
- Charging for Investigational Drugs Under an IND – Q&A (UCM351264)

For IDE applications:

- Frequently Asked Questions about Medical Devices (UCM127067)
- Significant Risk and Non-Significant Risk Medical Device Studies (UCM126418)

A study drug may also be taken from a research supply, which could be similar to the clinical supply but may be labeled as "not for human use," as it may not meet all of the manufacturing standards required for a therapeutic agent to be labeled as being from a clinical or commercial supply. Understanding the source of the product provided by the supporting company is an expectation of a sponsor or sponsor-investigator; therefore, it is key that the individual serving as sponsor ensures that they know the provenance of the drug. If the drug provided is from the clinical supply, an IND will most likely be required. If the drug provided is from the commercial supply, an IND may not be required depending on the drug's use in the study protocol.

When Is an IND Needed for a Dietary Supplement, Food, or Cosmetic Product?

There are times when the use of a drug, device, dietary supplement, or cosmetic product in a clinical investigation does not require an IND. The prescribing or use of a lawfully marketed product within the scope of practice of a licensed healthcare professional is considered the practice of medicine and is generally not regulated by the FDA and therefore does not require an IND (i.e., off-label use of a lawfully marketed drug in a single, patient-specific treatment plan). However, there are times when a product that is lawfully marketed as a dietary supplement, food, or cosmetic product may require an IND when used in certain clinical investigations.

Although it is often thought that the key differentiating factor between a drug and a dietary supplement is the intended purpose of the clinical investigation (i.e., whether the product being studied or marketed for the diagnosis, cure, mitigation, treatment, or prevention of disease), the definition of a drug also includes those compounds "intended to affect the structure or function of the body, without regard to whether the compound is intended to influence a disease process" (OFR 2010). An example of this is healthy individuals prescribed oral contraceptives to prevent pregnancy. Generally, an IND is required for a clinical investigation of a compound intended to alter the physiological response, study the mechanism of action, or study the metabolism of the compound. In this context, the compound is not considered a drug if it is a dietary supplement intended only to affect the structure or function of the body or food used only for taste, aroma, or nutritive value (i.e., not being used for a therapeutic purpose) (OFR 2019g; Nutrilab, Inc. v. Schweiker 1983). A dietary supplement is defined as "a product taken by mouth that is intended to supplement the diet and that contains one or more dietary ingredients" (NIH 1994). A clinical investigation of a dietary supplement would require an IND if the study were designed to evaluate a therapeutic purpose of the compound (i.e., the ability of Vitamin D to decrease time to wound healing) or was intended for use in the diagnosis, cure, mitigation, treatment, or prevention of disease. Similarly, a food product would require an IND and be considered a drug if it is "intended for use in the diagnosis, cure, mitigation, treatment, or prevention of disease" (OFR 2010). Authorized health claims, as per 21 CFR Part 101, may be made without the food becoming a drug (OFR 2019g).

A cosmetic is "(1) intended to be rubbed, poured, sprinkled, or sprayed on, introduced into, or otherwise applied to the human body or any part thereof for cleansing, beautifying, promoting attractiveness, or altering the appearance, and (2) articles intended for use as a component of any such articles; except that such term shall not include soap" (OFR 2010). Ingredients or products marketed as cosmetics require an IND if the ingredient is being studied "for use to affect the structure or function of the body or to prevent, treat, mitigate, cure, or diagnose a disease" (OFR 2019l; OFR 2019j). Even if the claim is related to the product's ability to cleanse or alter appearance, rather than a specific effect on the structure or function of the body, an IND is needed to conduct the clinical investigation. For example, a shampoo marketed with an antidandruff claim must comply with the requirements for both cosmetics and drugs; however, shampoo with the same ingredients that does not make the antidandruff claim may not.

KEY CONSIDERATIONS FOR A SPONSOR-INVESTIGATOR

When considering whether to take on the role of SI in an IISR study, a researcher must think critically about whether their skills and resources are appropriate for the project. Important factors for researchers to reflect upon include whether they have the regulatory and clinical experience and knowledge to take on the study, and if they have the time and desire to incorporate the additional responsibilities required into their day-to-day work. Awareness of available resources, both human and capital, is necessary, as is the availability of appropriate study populations. These details are important not only for the success of the research from a scientific standpoint, but also from a regulatory and legal perspective.

Principal Investigator Considerations

As a researcher contemplating the role of SI, one of the first things to reflect upon is whether the researcher has the experience, training, and knowledge needed to take on this work. This includes both clinical and regulatory experience, as the individual serving as study sponsor is responsible for understanding and adhering to regulations applicable to the study. If sponsorship is a potential part of the researcher's role, they must have the desire to take on the additional regulatory responsibilities, which go far above and beyond those expected of a PI. There are often additional financial costs and duties associated with the sponsorship role with which researchers should familiarize themselves.

One of the most important *resources* for an SI is time; this includes time devoted to regulatory oversight of the study, supervising the conduct of the study, and training study staff. This is frequently overlooked in the excitement of tackling a new research project. Researchers and clinicians planning IISR must think carefully about their other obligations and appointments, including travel. The sponsor is responsible for the conduct of the study, even *in absentia*. Ultimately, the sponsor or SI is held legally accountable for study conduct and supervision, even if delegated. To that end, it is crucial that anyone to whom oversight may be assigned is appropriately trained in how to comply with the study protocol and regulations.

Staffing Considerations

The success of any IISR study depends not only on the quality and quantity of study staff but also on their relative areas of expertise. While a study coordinator is able to serve as the point person managing collective efforts and steering the research unit, it is necessary to ensure that there are enough other members of the study staff available to move the project forward. Furthermore, it is up to the SI to determine what areas of expertise study staff might need to bring to the project and to ensure that they are adequately trained, with the appropriate competencies, licensures, and experience. This includes clinical, research/laboratory, and administrative members of the team.

Often overlooked in planning for an investigation is study staff turnover. Clinical trials may take a few years to complete, so anticipating the frequency of staff member departures is important to continual and coherent progress. In creating the study protocol and making decisions about staffing, the SI would do well to assess the required expertise needed for each study role, including learning curves and any additional training that may be needed for a new hire.

Legal, Financial, and Administrative Considerations

In the case of a study in which a company is providing funding and/or products for the investigation, legal, financial, and administrative factors should be thoroughly reviewed. From a legal perspective, the SI should check with their home institution to see whether they have the authority to sign any contracts with the company; if not, the researcher should work with their academic institution to have any legally binding documents signed by the appropriate parties. The sponsor is also expected to understand what the obligations stipulated in the contracts are, and whether there are any penalties for not following these requirements. Confidentiality regarding intellectual property (IP) should be clearly outlined between the company and the sponsor. This includes discussions of whether there is a patent for the product or whether the IP is proprietary and confidential. SIs should clarify whether any new ideas generated during the study are the property of the company and, if so, what benefits might return to the SI. All conflicts of interest need to be reviewed and appropriately disclosed by all parties.

Financially, budgets and payments from the company to the SI should be clearly outlined. In requesting funding, SIs should exercise due diligence in ensuring that they are asking for funding and products with FMV in mind. We recommend that SIs ask for clarification on any budget or product provision area where there is ambiguity or doubt; this is due in part to the fact that it is nearly impossible to return to the company to ask for additional supplies or monies. This includes inquiring about payment paradigms (i.e., per subject enrolled) and the timing of payments (i.e., upon completion of the study by each subject). It is also up to the SI to determine whether the payments to support the study should be made to their institution or to them.

Lastly, when preparing to undertake a project as an SI, a researcher should work closely with their institution's grants management office. This office has the expertise to ensure that all of the logistics, including payment schedules and/or invoicing, are handled correctly. It is worthwhile to reach out to the grants management team early in the process, as they can help ensure that the administrative components of a study conducted under a grant are completed correctly and at the right time, helping to avoid waste or redundant work.

Facility Considerations

Even if they are experts in the clinical and regulatory areas of the project, an SI should only undertake research for which their facilities are equipped. For example, if the investigator is conducting a study that requires an overnight stay, but their facility does not have an adequate number of beds, the environment would not be appropriate for that particular investigation. The term "facilities" refers not only to the physical space but also to the necessary equipment and appropriate technologists to run said equipment. A study involving the use of an X-ray machine would require a technologist to be available to run it, and a radiologist to read the resulting images; if the X-ray machine is available, but no one has the time or training to use it, then the facility may not be properly equipped to carry out that trial. Not only do the technologists need to be at the institution, but they need to have the flexibility to be available when needed (i.e., having phlebotomists on hand for timed blood draws). One of the challenges is that the technologists and nurses whom one might ask to be involved in the study have a full load of clinical duties; therefore, the study activities may need to be timed in conjunction with the associated clinic(s). A facility requirement that is frequently overlooked by SIs is the need for adequate and secure storage for the investigational product(s) and study records. All components of a clinical investigation need to be stored separately from day-to-day clinical products and records; even if the investigational drug or device is already commercially available, it needs to be stored, labeled, and documented separately from non-investigational daily clinical supplies. As each product needs to be tracked and monitored, storage areas for investigational products should have limited accessibility to non-study staff. A carefully kept contemporaneous study drug/device use log must be maintained as well to document when a product is removed from storage and used, as well as which patient received it (often identifying patients by study identification number in the log). Records associated with the clinical investigation need to be stored separately from non-research clinical care records, including case report forms. If multiple studies are being conducted at the same site, SIs must ensure that each study's products, case reports, and documentation are maintained in separate storage areas. Any electronic systems used must comply with the appropriate regulations, including those that apply to document control and quality systems (OFR 2019a). Researchers should consult their institution's information technology quality office for assistance in meeting the appropriate safety and encryption requirements for electronic systems.

When determining whether a laboratory has the capacity to support a clinical investigation, SIs should assess several characteristics of the laboratory. It is the responsibility of the SI to determine whether the laboratory has sufficient capabilities to complete necessary assays and testing; this includes having appropriate, well-maintained equipment and staff trained to use it, ensuring that the equipment and technologists will be available when needed, and checking that the equipment can accomplish the necessary tasks with the level of specificity and accuracy needed. If the SI determines that the equipment or services available in-house will not be adequate for the study, they may want to consider outsourcing the testing to contract labs. This should be assessed early in the study planning, as these laboratories often cost more to use and should therefore be incorporated into proposed study budgets. When vetting such outside testing facilities, it is the SI's job to identify reputable, vetted labs with appropriate licensure. Storage within the laboratory should also be considered, including temperature, humidity, and other factors; conditions need to be appropriately monitored and documented throughout the investigation. If a refrigerator or freezer is needed for the investigational products, it should be validated and dedicated to storing only study products. If specialized diagnostic equipment or testing is needed, such as laboratory-developed tests (LDTs), SIs should familiarize themselves with the appropriate guidances and documents (USFDA, CDRH, CBER 2014).

Protocol Considerations

As a researcher develops their protocol for an IISR study, they must take the time to evaluate both internal and external factors that have the potential to impact the study. Internal factors include the researcher's previous experience with similar protocols, drugs, and devices. If the investigator does not have prior experience with the proposed protocols or studies, or has not previously used similar drugs and/or devices, they must find help, be trained, or reconsider conducting the study in question. This includes having familiarity with research techniques such as blinding or cross-over design, both of which require unique documentation and regulatory processes compared to single-arm, open-label studies.

External factors include subject availability, other studies being conducted both intra- and extra-murally, facility appropriateness, and availability of staff to support the study. When determining whether there is an adequate number of potential study subjects in the area, researchers should ask themselves if these are the kind of patients who would want to be part of such a study, and if these patients are reliable (i.e., unlikely to be lost to follow-up). Statisticians should be consulted as early as possible to design a study with adequate enrollment targets to account for drop-out and/or loss to follow-up. Another often-overlooked task is to review the other trials going on both at their own institution and at institutions in the surrounding area, as other studies may have similar inclusion/exclusion criteria and therefore may be in direct competition with the researcher's proposed trial. Finally, as discussed earlier, the availability of facilities, equipment, and personnel must be considered when developing an IISR study protocol.

Compliance History

As part of the assessment a supporting company conducts when evaluating the researcher's qualifications, study protocol, and plans, it is highly recommended that the company review FDA inspection histories of participating study investigators, facilities, institutions, and IRBs. Key questions to ask during such reviews are whether any of the aforementioned parties have ever been inspected and/or received a Warning Letter; this is especially critical for the IRB, as they will be the ones to approve the study on the institutional level. If inspections have taken place and a Notice of Observation (Form FDA 483) has been issued, the supporting company should follow up with the SI to determine if the FDA district office was consulted to address any concerns and avoid the issuance of a Warning Letter. Establishment Inspection Reports (EIRs) resulting from FDA inspections are sent to the FDA's headquarters to determine if a Warning Letter should be issued. Most of these reports are publicly available through FDA's databases; however, the EIRs must be requested through the Freedom of Information Act (FOIA) (USFDA 2013, 2021).

Equally important, if not more so, is determining whether any of the parties potentially involved in the clinical investigation are on the FDA's disqualified investigator or restricted investigator lists. If so, the time frame for disqualification or restriction should be checked through the FDA website. Note that just because an SI, institution, or IRB has received a Warning Letter does not mean that they cannot participate in the research. If the investigator's period of disqualification/restriction has expired, the investigator may participate. If a potential investigator is listed on an FDA debarment list, they should immediately be taken out of consideration for study support.

Additional Subject Considerations

In addition to determining whether patients are available who meet the necessary inclusion/exclusion criteria, SIs ought to consider the sociodemographic characteristics of their potential subjects. For example, some individuals may not own a car and may therefore be traveling to the study site via public transportation; if no buses or subway stations are in the vicinity of the clinic, those individuals may be less likely to enroll in the study or may be likelier to drop out or be lost to follow-up. Similarly, if the study site aims to draw subjects from rural areas where public transit is not available, that may impede enrollment and adherence to protocol. SIs may want to consider earmarking funds to offset study participants' transit costs to reduce the possibility of travel as a barrier to follow-up. For those living far away, SIs may want to work with the subject's primary care physician to conduct follow-up locally.

Beyond travel, the general reliability of potential subjects is also an important factor. Study participants must be trusted to provide accurate health data and self-reported outcomes, which may be compromised due to either a lack of knowledge or a desire to fulfill social expectations. Some subjects may be *professional research subjects*, meaning that they enroll in multiple studies simultaneously as a way of making a living. It is critical in an investigational study that subjects are completely transparent regarding their enrollment in any other studies to ensure that there are no adverse events or drug-drug interactions that could be harmful to the subject or adversely impact the validity of the study.

IRB Considerations

The SI must evaluate the IRB itself, including whether the IRB to be used is compliant with the necessary regulations and is accredited by the Association for the Accreditation of Human Research Programs (AAHRPP). Regulations with which the IRB must comply include those related to human subject protection, investigator financial disclosure, and IRB; these include regulations in the Code of Federal Regulations and the Office of Human Research Protections (OHRP) and obtaining a Federal-Wide Assurance (FWA) (OFR 2019b, 2019e, 2019f). IRBs may be local (i.e., within the academic institution) or central (outside of the academic institution); review should be budgeted and paid for by the investigator.

Investigators should also inquire about the frequency with which the IRB meetings are held and the typical time for review. It is useful to inquire about the ratio of approvals to non-approvals, and how many stipulations and/or iterations are typically needed before approval. If IISR involves vulnerable populations, radiation, or biohazards, the researcher should ensure that their IRB has the necessary special review boards. If not, it may be necessary to work with a contract IRB to have it reviewed appropriately.

Open Payments Considerations

Open Payments is a transparency program created by section 6002 (Physician Payment Sunshine Act) of the Affordable Care Act (ACA), which increases public awareness of financial relationships between the healthcare industry (such as medical device manufacturers and pharmaceutical companies) and healthcare providers (physicians, pharmacists, teaching hospitals, etc.) (111th Congress of the USA 2010). Medical residents are excluded from the definition of physician for the purpose of this program. Drug, device, biological, and medical supply manufacturers are required to report payments or transfers of value they make to providers. The Centers for Medicare & Medicaid Services (CMS) collects this data and makes it publicly available and searchable online at https://www.cms.gov/openpayments (CMS 2017). Of note, providers do not have to treat Medicare or Medicaid patients to be subject to the regulations.

Collaboration among physicians, teaching hospitals, and health care manufacturing companies contributes to the design and delivery of many life-saving drugs and devices. However, sometimes these payments can also introduce conflicts of interest. Open payments don't identify which financial relationships are beneficial or which may cause conflicts of interest. Payments or transfers of value include, but are not limited to, consulting fees, speaker fees, honoraria, gifts, entertainment, food or beverage, travel, education, research activities, charitable donations, royalties, licenses, grants, or investments.

FDA Inspections of Clinical Investigators

In 1977, the Food and Drug Administration (FDA) established the Bioresearch Monitoring (BIMO) program to ensure the protection of the rights, safety, and welfare of human research subjects involved in FDA-regulated clinical studies; to verify the accuracy and reliability of clinical study data submitted to the FDA in support of research or marketing applications; and to assess compliance with statutory requirements and the FDA's regulations governing the conduct of clinical studies, including those for informed consent and ethical review (USFDA 2020). The FDA's Office of Regulatory Affairs (ORA) conducts inspections of clinical investigators, sponsors, sponsor-investigators, monitors, contract research organizations (CROs), institutional review boards (IRBs), and nonclinical (animal) laboratories (USFDA 2020).

The goal of an FDA inspection of a clinical investigator or investigational site is to confirm the validity and reliability of data submitted to the FDA, assess the investigator's conduct of the trial, ensure appropriate protection of human subjects, and confirm that the study is in compliance with applicable regulations (USFDA 2006). Inspections may be initiated for a variety of reasons, including a result of a complaint regarding a specific site, a concern from a sponsor or FDA review division, termination of a site, or in response to a specific public health concern. An inspection may help the FDA determine if a study supporting an IND, NDA, BLA, PMA, 510(k), or other drug or device application was well-designed, conducted according to Good Clinical Practice (GCP), and if the results reported are accurate and applicable human subject protections were followed (USFDA 2020). This may be especially helpful in circumstances where the study was not conducted in the United States. For instance, a premarket approval (PMA) application may be supported by an international clinical study not conducted under an investigational device exemption (IDE). The FDA may inspect the investigational site to ensure all of the aforementioned criteria were followed despite the lack of an IDE and previous FDA oversight.

FDA may or may not notify an investigator of an inspection prior to the start of the inspection. It is a prohibited act under sections 301(e) and 301(f) of the FDCA to refuse to permit FDA entry or inspection or to refuse to permit access to or copying of certain specified records. At the start of an inspection, the FDA inspector will provide his/her credentials and a Notice of Inspection (Form FDA 482) to the appropriate study staff or clinical investigator (USC 2010, OFR 2019n, 2019q). The FDA is authorized to access, verify, and copy any reports or records related to the study, including protected health information, if FDA believes it is relevant to the inspection. In addition to physically inspecting records, FDA may also speak with individuals to confirm compliance with all regulatory requirements, including but not limited to following the study protocol, completing all required IRB processes, following appropriate informed consent procedures, reporting any potential conflicts of interest, appropriately delegating clinical trial monitoring responsibilities, adequately training study staff, and supervising the study by the investigator (111th Congress of the USA 2010; USFDA 2006, 2020; USC 2010).

After the inspection is completed, the FDA investigator will conduct an exit interview with the clinical investigator or the appropriate study staff. The FDA investigator will review observations and findings from the inspection that, in their judgment, may be considered deficiencies. Common deficiencies identified during inspections include protocol deviations, inadequate recordkeeping, failure to supervise, and inadequate accountability for the investigational product. The FDA investigator will issue a written Inspectional Observations (Form FDA 483) to the clinical investigator or study staff and then send the inspection report to the FDA District Office. The Establishment Inspection Report (EIR), Form FDA 483, along with any other relevant records or materials collected during the inspection, will ultimately go to the appropriate FDA Center for review and to determine if a Warning Letter should be issued. Investigators *may* respond to Form FDA 483, if desired, to the inspector during the exit interview or in writing to the FDA District Office (USC 2010). Even though there is no regulatory requirement to respond to Form FDA 483, it is encouraged by the FDA and generally beneficial for the clinical investigator to do so. An effective response to Form FDA 483 may mitigate further compliance action, demonstrate to the FDA an understanding of the observations, a commitment to correct the deficiencies, and voluntary compliance with the regulations, and establish credibility. Elements of an effective response to Form FDA 483 are listed in Box 2.5.

BOX 2.5 Elements of an Effective Response to Form FDA 483

The following are important to consider when responding to Form FDA 483:

- Include a commitment/statement from senior leadership.
- Address each observation separately.
- Note whether you agree or disagree with the observation.
- Provide corrective action accomplished and/or planned; tell the FDA the plan.
- Provide time frames for corrections.
- Provide method(s) of verification and/or monitoring for corrections.
- Consider submitting documentation of corrections where reasonable and feasible.

The result from the FDA review will likely be one of four scenarios: no deficiencies were found, deficiencies identified that do not meet the threshold of regulatory significance for a Warning Letter, serious deviations or deficiencies that warrant a Warning Letter, or a Notice of Initiation of Disqualification Proceedings and Opportunity to Explain (NIDPOE) (USC 2010; OFR 2019q). A written letter detailing a lack of deficiencies may not be sent to a clinical investigator. Deficiencies that do not lead to a Warning Letter often still require a written response from the clinical investigator. If deficiencies identified in a Warning Letter are not voluntarily and adequately addressed by the investigator within 15 working days of the investigator's receipt of the Warning Letter, there may be additional consequences for the investigator. Failure to adequately and promptly explain the violations noted in the Warning Letter may result in regulatory action without further notice.

Lastly, a NIDPOE allows an investigator an opportunity to explain the results of the inspection, either verbally or in writing. The FDA may then accept the investigator's explanation and identify appropriate corrective action or issue a Notice of Opportunity for Hearing (NOOH) and move to disqualify the investigator from conducting investigational drug or device research. All Warning Letters and NIDPOE are sent to the study sponsor and affiliated IRB. Warning Letters, NIDPOE, and NOOH are publicly available on the FDA's website (USFDA 2021b, 2021c). Forms FDA 483 are available through the FOIA and may be requested on the FDA's website (USFDA 2018).

Record Retention

Study records for investigational drug research should be retained for at least two years after the date a marketing application is *approved* for the drug and indication supported by the study (OFR 2019m). If a marketing application will not be pursued, study records should be retained for at least two years after the date the study is discontinued and FDA is notified of the study discontinuation (OFR 2019m). Similarly, study records for investigational device research should be retained for two years after the investigation is completed, terminated, or the records are no longer required to support a premarket approval application, whichever is later (OFR 2019p).

RISK CONSIDERATIONS AND COMMON MISTAKES

Another form of risk management that needs to be considered is that of public perception of the study. If the research involves controversial materials or technologies, such as embryonic stem cells, media interest may need to be managed by an experienced public relations person. If the work has the potential to bring about debate, the investigator may want to consider keeping the study low-profile until its completion. The perceived value of the study may need to be communicated to the patients, institutions, and public to generate *buy-in* from necessary stakeholders.

New SIs frequently overlook a few key items when launching their investigations. One of these common mistakes is inadequate project support, including too small of a budget and not establishing clear expectations regarding IP and data ownership. Inadequate study support is also cited as a frequent issue, particularly concerning protocol, statistics, scientific review, subject availability, and personnel. A lack of regulatory support may also be an issue. Almost all of these can be anticipated and prevented if the considerations detailed earlier in this chapter are included in the planning stages of IISR.

It is recommended that researchers take advantage of resources that may already be available to them at their institution to help during the early phases of the project. For example, if help is needed with regard to the drug or device being tested, an SI might inquire as to whether the company with whom they are partnering has specialists who might be able to assist; note, however, that this assistance may have costs associated with it or the company may have a policy not to provide any support to the SI other than financial and product support. If support is needed in study design, statistics, or regulatory support, the SI could make use of institutional resources, such as statistical cores, offices of regulatory research support, or a center for clinical and translational sciences. Additional regulatory support can be found on the FDA's website, through a hired regulatory consultant, or through the pre-submission programs offered by the FDA (USFDA 2009, 2014).

PARTING THOUGHTS

IISR plays a key role in translating findings from the laboratory to the clinic and population. In this chapter, we have defined the role of the SI, provided information on the regulatory role of the SI, and detailed some key factors a potential sponsor-investigator must consider when beginning an investigator-driven project. Should a researcher decide to serve as an SI, careful planning can mean that a study's results could be brought to patients more efficiently. Furthermore, a well thought-out clinical study has the potential to not only impact clinical care in the United States but public health worldwide.

REFERENCES

111th Congress of the United States of America (USA). March 22, 2010. "The Patient Protection and Affordable Care Act: H.R. 3590, 111th Congress." https://www.govtrack.us/congress/bills/111/hr3590/text.

Center for Medicare & Medicaid Services (CMS). November 8, 2017. "What is Open Payments?" https://www.cms.gov/openpayments.

National Institutes of Health (NIH), Office of Dietary Supplements. October 25, 1994. "Dietary Supplement Health and Education Act of 1994: Public Law 103-417, 103rd Congress." https://ods.od.nih.gov/About/DSHEA_Wording.aspx.

Nutrilab, Inc. v. Schweiker, 713 F.2d 335 (7th Cir. 1983).

Office of the Federal Register (OFR), National Archives and Records Administration. 2010. "Code of Federal Regulations." Title 21 Food and Drugs, Chapter 9, Part 321 - Definitions. https://www.govinfo.gov/content/pkg/USCODE-2010-title21/pdf/USCODE-2010-title21-chap9-subchapII-sec321.pdf.

Office of the Federal Register (OFR), National Archives and Records Administration. April 1, 2019a. "Code of Federal Regulations." Title 21 Food and Drugs, Volume 1, Part 11 - Electronic records; electronic signatures. https://www.govinfo.gov/content/pkg/CFR-2020-title21-vol1/pdf/CFR-2020-title21-vol1.pdf.

Office of the Federal Register (OFR), National Archives and Records Administration. April 1, 2019b. "Code of Federal Regulations." Title 21 Food and Drugs, Volume 1, Part 50 - Protection of human subjects. https://www.govinfo.gov/content/pkg/CFR-2020-title21-vol1/pdf/CFR-2020-title21-vol1.pdf.

Office of the Federal Register (OFR), National Archives and Records Administration. April 1, 2019c. "Code of Federal Regulations." Title 21 Food and Drugs, Volume 1, Part 50.25 - Protection of human subjects: Elements of informed consent. https://www.govinfo.gov/content/pkg/CFR-2020-title21-vol1/pdf/CFR-2020-title21-vol1.pdf.

Office of the Federal Register (OFR), National Archives and Records Administration. April 1, 2019d. "Code of Federal Regulations." Title 21 Food and Drugs, Volume 1, Part 50.3 - Protection of human subjects: Definitions. https://www.govinfo.gov/content/pkg/CFR-2020-title21-vol1/pdf/CFR-2020-title21-vol1.pdf.

Office of the Federal Register (OFR), National Archives and Records Administration. April 1, 2019e. "Code of Federal Regulations." Title 21 Food and Drugs, Volume 1, Part 54 - Financial disclosure by clinical investigators. https://www.govinfo.gov/content/pkg/CFR-2020-title21-vol1/pdf/CFR-2020-title21-vol1.pdf.

Office of the Federal Register (OFR), National Archives and Records Administration. April 1, 2019f. "Code of Federal Regulations." Title 21 Food and Drugs, Volume 1, Part 56 - Institutional Review Boards. https://www.govinfo.gov/content/pkg/CFR-2020-title21-vol1/pdf/CFR-2020-title21-vol1.pdf.

Office of the Federal Register (OFR), National Archives and Records Administration. April 1, 2019g. "Code of Federal Regulations." Title 21 Food and Drugs, Volume 2, Part 101 - Food labeling. https://www.govinfo.gov/content/pkg/CFR-2020-title21-vol2/pdf/CFR-2020-title21-vol2.pdf.

Office of the Federal Register (OFR), National Archives and Records Administration. April 1, 2019h. "Code of Federal Regulations." Title 21 Food and Drugs, Volume 4, Part 210 - Current good manufacturing practice in manufacturing, processing, packing, or holding of drugs; general. https://www.govinfo.gov/content/pkg/CFR-2020-title21-vol4/pdf/CFR-2020-title21-vol4.pdf.

Office of the Federal Register (OFR), National Archives and Records Administration. April 1, 2019i. "Code of Federal Regulations." Title 21 Food and Drugs, Volume 4, Part 211 - Current good manufacturing practice for finished pharmaceuticals. https://www.govinfo.gov/content/pkg/CFR-2020-title21-vol4/pdf/CFR-2020-title21-vol4.pdf.

Office of the Federal Register (OFR), National Archives and Records Administration. April 1, 2019j. "Code of Federal Regulations." Title 21 Food and Drugs, Volume 5, Part 312.2 - Investigational new drug application: Applicability. https://www.govinfo.gov/content/pkg/CFR-2020-title21-vol5/pdf/CFR-2020-title21-vol5.pdf.

Office of the Federal Register (OFR), National Archives and Records Administration. April 1, 2019k. "Code of Federal Regulations." Title 21 Food and Drugs, Volume 5, Part 312.23 - Investigational new drug application: IND content and format. https://www.govinfo.gov/content/pkg/CFR-2020-title21-vol5/pdf/CFR-2020-title21-vol5.pdf.

Office of the Federal Register (OFR), National Archives and Records Administration. April 1, 2019l. "Code of Federal Regulations." Title 21 Food and Drugs, Volume 5, Part 312.3 - Investigational new drug application: Definitions and interpretations. https://www.govinfo.gov/content/pkg/CFR-2020-title21-vol5/pdf/CFR-2020-title21-vol5.pdf.

Office of the Federal Register (OFR), National Archives and Records Administration. April 1, 2019m. "Code of Federal Regulations." Title 21 Food and Drugs, Volume 5, Part 312.62 - Investigational new drug application: Investigator recordkeeping and record retention. https://www.govinfo.gov/content/pkg/CFR-2020-title21-vol5/pdf/CFR-2020-title21-vol5.pdf.

Office of the Federal Register (OFR), National Archives and Records Administration. April 1, 2019n. "Code of Federal Regulations." Title 21 Food and Drugs, Volume 5, Part 312.68 - Investigational new drug application: Inspection of investigator's records and reports. https://www.govinfo.gov/content/pkg/CFR-2020-title21-vol5/pdf/CFR-2020-title21-vol5.pdf.

Office of the Federal Register (OFR), National Archives and Records Administration. April 1, 2019o. "Code of Federal Regulations." Title 21 Food and Drugs, Volume 5, Part 312.8 - Investigational new drug application: Charging for investigational drugs under an IND. https://www.govinfo.gov/content/pkg/CFR-2020-title21-vol5/pdf/CFR-2020-title21-vol5.pdf.

Office of the Federal Register (OFR), National Archives and Records Administration. April 1, 2019p. "Code of Federal Regulations." Title 21 Food and Drugs, Volume 8, Part 812.140 - Investigational device exemptions: Records. https://www.govinfo.gov/content/pkg/CFR-2020-title21-vol8/pdf/CFR-2020-title21-vol8.pdf.

Office of the Federal Register (OFR), National Archives and Records Administration. April 1, 2019q. "Code of Federal Regulations." Title 21 Food and Drugs, Volume 8, Part 812.145 - Investigational device exemptions: Inspections. https://www.govinfo.gov/content/pkg/CFR-2020-title21-vol8/pdf/CFR-2020-title21-vol8.pdf.

Office of the Federal Register (OFR), National Archives and Records Administration. April 1, 2019r. "Code of Federal Regulations." Title 21 Food and Drugs, Volume 8, Part 812.3 - Investigational device exemptions. https://www.govinfo.gov/content/pkg/CFR-2020-title21-vol8/pdf/CFR-2020-title21-vol8.pdf.

United States Code (USC). 2010. "Code of Federal Regulations." Title 21 Food and Drugs, Chapter 9: Federal Food, Drug, and Cosmetic Act, Part 374 - Inspection. https://www.govinfo.gov/content/pkg/USCODE-2010-title21/html/USCODE-2010-title21-chap9-subchapVII-partA-sec374.htm.

United States Food and Drug Administration (USFDA). 2009. "Guidance for Industry: Formal Meetings between the FDA and Sponsors or Applicants."

United States Food and Drug Administration (USFDA). June 28, 2013. "FDA Freedom of Information Act (FOIA): FDA FOIA On-Line Request Submission Information." https://www.accessdata.fda.gov/scripts/foi/FOIRequest/requestinfo.cfm.

United States Food and Drug Administration (USFDA). 2014. "Guidance for Industry and Food and Drug Administration Staff: Requests for Feedback on Medical Device Submissions: The Pre-Submission Program and Meetings with Food and Drug Administration Staff."

United States Food and Drug Administration (USFDA). July 2017. "Form FDA 3574 Supplement – Form Instructions." https://www.fda.gov/media/86100/download.

United States Food and Drug Administration (USFDA). August 20, 2018. "Freedom of Information." https://www.fda.gov/regulatory-information/freedom-information.

United States Food and Drug Administration (USFDA). March 2019. "Form FDA 1571 Supplement – Form Instructions." https://www.fda.gov/media/77596/download.

United States Food and Drug Administration (USFDA). July 22, 2020. "FDA Compliance Program Clinical Investigators and Sponsor-Investigators 7348.811: CHAPTER 48-BIORESEARCH MONITORING." https://www.fda.gov/media/75927/download.

United States Food and Drug Administration (USFDA). 2021a. "Inspection Classification Database Search." https://www.accessdata.fda.gov/scripts/inspsearch/.

United States Food and Drug Administration (USFDA). March 2, 2021b. "Clinical Investigators - Disqualification Proceedings." https://www.accessdata.fda.gov/scripts/SDA/sdNavigation.cfm?sd=clinicalinvestigatorsdisqualificationproceedings&previewMode=true&displayAll=true.

United States Food and Drug Administration (USFDA). March 28, 2021c. "Warning Letters." https://www.fda.gov/ICECI/EnforcementActions/WarningLetters/default.htm.

United States Food and Drug Administration (USFDA), Center for Devices and Radiological Health (CDRH). January 2006. "Information Sheet Guidance For IRBs, Clinical Investigators, and Sponsors: Significant Risk and Nonsignificant Risk Medical Device Studies." http://www.fda.gov/downloads/RegulatoryInformation/Guidances/UCM126418.pdf.

United States Food and Drug Administration (USFDA), Center for Drug Evaluation and Research (CDER). March 1991. "Guidance for Industry: Guideline on the Preparation of Investigational New Drug Products (Human and Animal)." https://www.fda.gov/media/71017/download.

United States Food and Drug Administration (USFDA), Center for Devices and Radiological Health (CDRH), Center for Biologics Evaluation and Research (CBER). December 5, 2017. "FDA Categorization of Investigational Device Exemption (IDE) Devices to Assist the Centers for Medicare and Medicaid Services (CMS) with Coverage Decisions: Guidance for Sponsors, Clinical Investigators, Industry, Institutional Review Boards, and Food and Drug Administration Staff." https://www.fda.gov/media/98578/download.

United States Food and Drug Administration (USFDA), Center for Drug Evaluation and Research (CDER), Center for Biologics Evaluation and Research (CBER), Center for Devices and Radiological Health (CDRH), Office of Good Clinical Practice (OGCP), Office of Regulatory Affairs (ORA). August 2013. "Guidance for Industry: Oversight of Clinical Investigations – A Risk-Based Approach to Monitoring." https://www.fda.gov/media/116754/download.

United States Food and Drug Administration (USFDA), Center for Drug Evaluation and Research (CDER), Center for Biologics Evaluation and Research (CBER), Office of Regulatory Affairs (ORA). July 2008. "Guidance for Industry: CGMP for Phase 1 Investigational Drugs." https://www.gmp-compliance.org/files/guidemgr/indcgmp.pdf.

United States Food and Drug Administration (USFDA), Center for Devices and Radiological Health (CDRH), Office of *In Vitro* Diagnostics and Radiological Health, Center for Biologics Evaluation and Research (CBER). October 3, 2014. "Draft Guidance for Industry, Food and Drug Administration Staff, and Clinical Laboratories: FDA Notification and Medical Device Reporting for Laboratory Developed Tests (LDTs)." https://www.fda.gov/media/89837/download.

10 Medical Countermeasures for Public Health Emergency Preparedness and Response in the United States

A Regulatory Perspective

Angela Bellos, Jack Russell, and Doris Snow

INTRODUCTION

On January 31, 2020, under Section 319 of the Public Health Service (PHS) Act, a nationwide public health emergency (PHE) was declared by the U.S. Secretary of Health and Human Services (HHS) due to confirmed cases of the 2019 novel coronavirus (Office of the Assistant Secretary for Preparedness and Response [ASPR] 2021), as the virus had been provisionally called prior to its designation as the severe acute respiratory syndrome coronavirus 2 (SARS-CoV-2) (Coronaviridae Study Group of the International Committee on Taxonomy of Viruses 2020), the etiologic agent of coronavirus disease 2019 (COVID-19). On February 4, 2020, under Section 564 of the Federal Food, Drug, and Cosmetic (FD&C) Act, the Secretary of HHS also determined the PHE had "significant potential to affect national security or the health and security of United States citizens living abroad," so he declared circumstances existed to justify EUA of certain MCMs (Department of HHS 2020d). In addition to the PHE, on March 13, 2020, a national emergency concerning COVID-19, which was determined to have begun March 1, 2020, was declared under Sections 201 and 301 of the National Emergencies Act by Proclamation 9994 (Executive Office of the President [EOP] 2020). Prior to the COVID-19 pandemic, other emergencies declared in the U.S. that were caused by infectious diseases included a national emergency concerning the H1N1 influenza pandemic that was renewed twice in 2009 (EOP 2009) and a nationwide PHE for the same that was renewed in 2010, as well as

 DOI: 10.1201/9781315117911-10

a PHE in Puerto Rico in 2016 for Zika virus that was also renewed in 2017 (Office of the ASPR 2023). The national responses to these emergencies have been the subject of intense debate; however, the importance of improved approaches to rapidly and flexibly respond with effective countermeasures to safeguard the health and welfare of the public when the next infectious disease threat strikes is indisputable. A secure national supply of medicines produced using innovative manufacturing technologies will be an integral component of preparedness for the next pandemic.

Medical products that can counter infectious diseases and other threats are called medical countermeasures (MCMs). The overarching objective of this chapter is to provide an overview of MCMs for PHE preparedness and response in the United States (U.S.), including MCM development stages, authorities, legislation, regulations, and guidance. The topic is discussed primarily in the context of the COVID-19 pandemic and other events caused by biological threats in recent U.S. history, such as H1N1 influenza and Zika virus. The CDC estimated approximately 12,500 deaths in the U.S. were attributable to H1N1 influenza in the 2009 pandemic (Shrestha et al. 2011). By contrast, Zika virus infections are rarely fatal, with five deaths reported in Puerto Rico in the 2016–2017 outbreak; however, dozens of cases of birth defects and Guillain-Barré syndrome were also reported (Rodríguez-Díaz et al. 2018). The burden of COVID-19 has been staggering. From February to December 2020, an estimated 70 million symptomatic infections and 4 million hospitalizations occurred in the U.S. (CDC 2021a), with approximately 550,000 deaths reported to the CDC a year after the national emergency was declared (CDC 2021b). These PHEs testify to the urgency of rapid MCM development, fielding, and deployment when the nation faces an emerging infectious disease threat. Factoring in other components of disease burden and health disparities would throw into even starker relief the paramount importance of a comprehensive, coordinated, whole-of-government approach to national pandemic preparedness and response efforts to preserve public health and security.

In this context, this chapter first provides an overview of MCMs, with an emphasis on biological products and other drugs indicated for diseases and conditions caused by biological agents and toxins, and the authorities for MCM-related development activities and use. Thereafter, it delves into the central role played by the U.S. Food and Drug Administration (FDA) in fostering MCM development and monitoring the performance of deployed MCMs as part of its mission to protect public health. It also covers key regulations and guidance pertinent to MCM development, such as those focused on Emergency Use Authorization (EUA), Expanded Access, Orphan Designation, Expedited Programs, and the Animal Rule. It then describes the roles played by the FDA and other authorities within the Public Health Emergency Medical Countermeasures Enterprise (PHEMCE) in the development and lifecycle management of MCMs. Finally, it summarizes challenges and opportunities for MCM development in the U.S. regulatory space and offers an outlook for PHE preparedness and response with MCMs.

MEDICAL COUNTERMEASURES

As the COVID-19 pandemic has revealed, key to combating emerging or wholly novel infectious disease threats is the capability to respond in an operationally relevant timeframe with safe and effective MCMs, which are pharmaceutical or non-pharmaceutical medical interventions that can diagnose, prevent, mitigate, or treat infections or other diseases, conditions, or adverse health effects caused by chemical, biological, radiological, and nuclear (CBRN) threats, emerging infectious diseases, or natural disasters (PHEMCE 2017; CDC 2020a). MCMs may be approved or unapproved products that have been authorized for use. Qualified MCMs are more narrowly defined in Section 319F–1(a)(2) of the Public Health Service (PHS) Act. MCMs may be drugs, including biological products such as vaccines, antibodies, and antitoxins, or other drugs such as antibiotics and anticonvulsants; medical devices, such as *in vitro* diagnostics, personal protective equipment (PPE), decontamination systems, and ventilators; or any other medical product used as a countermeasure against biological threats like viruses, chemical threats like nerve gas, or radiologic or nuclear threats like a nuclear blast or reactor accident.

Approval, conditional approval, licensure, or clearance may be sought for MCMs, depending on their regulatory classification, but more expeditious pathways to make these products available to the intended users during a PHE include EUA, which may be issued for an unapproved product or for an approved product indicated for another use, or Expanded Access (also known as "compassionate use"), in which patients with a serious or life-threatening disease can access an investigational product (typically a therapeutic) outside a clinical trial, described in greater detail in the sections entitled "Emergency Use Authorization" and "Expanded Access," respectively. Timelines from initial review to EUA or Expanded Access may be days or weeks compared with months or years (potentially a decade or more) for approval. For example, COVID-19 vaccines were made available via EUA less than a year after the COVID-19 PHE was declared, with full approval approximately a year and a half after the declaration (FDA 2021l); the fastest any vaccine had ever been approved before then was four years for the mumps vaccine (Ball 2021). The MCMs deployed during the COVID-19 pandemic illustrate the wide spectrum of MCM types that may be required during a PHE and the mechanisms by which they can be rapidly made available to the public.

During the COVID-19 pandemic, the FDA authorized the use of or Expanded Access to an unprecedented number of MCMs (FDA 2021a). Over 400 *in vitro* diagnostics have been issued EUAs, the first of which was a real-time polymerase chain reaction diagnostic panel developed by the CDC that was authorized less than a week after the PHE was declared in early February 2020 (FDA 2023a), a testament to how expeditiously the FDA can facilitate MCM availability in a crisis situation. After EUAs for *in vitro* diagnostics, EUAs for other devices soon followed, with the first in late March 2020 for certain ventilators, ventilator tubing connectors, and ventilator accessories. In addition to ventilators (over 80 of which have been authorized for use), the FDA has also issued EUAs for over 60 other devices, including PPE like masks and protective barriers, predictive screening devices, remote patient monitors, and decontamination systems. Several such EUAs were revoked within a few months of issuance when, based on emerging scientific data, the known and potential benefits did not continue to outweigh the known and potential risks.

The first EUAs for COVID-19 drugs included antivirals such as chloroquine phosphate and hydroxychloroquine sulfate for the treatment of COVID-19 (March 2020) and remdesivir (Veklury®; Gilead Sciences, Inc.) for certain hospitalized COVID-19 patients (May 2020) (FDA 2022a). Several other EUAs followed, such as COVID-19 convalescent plasma for hospitalized COVID-19 patients (August 2020) and Eli Lilly's bamlanivimab for certain patients with mild-to-moderate COVID-19 (November 2020). Vaccines were the last drug product type to be issued EUAs, with the Pfizer-BioNTech COVID-19 vaccine (Comirnaty®) and Moderna COVID-19 vaccine (Spikevax®) (both original monovalent vaccines) authorized in December 2020 (FDA 2022b, 2022c). Approximately one year after the national emergency was declared, over 30 studies listed in clinicaltrials.gov involved Expanded Access for COVID-19 patients, such as remdesivir, the Johnson & Johnson's COVID-19 vaccine, COVID-19 convalescent plasma, mesenchymal stromal cells, and monoclonal antibodies (U.S. National Library of Medicine 2021).

Biological Threats

Most of the biological threats in recent U.S. history have been due to viruses, such as SARS-CoV-2, although bacteria, fungi, and their non-living products may be threats of biodefense concern (Calder and Bland 2015). The threat posed by live agents and the corresponding need for MCMs is based on a risk assessment of the pathogenicity, virulence, lethality, infectivity, and transmissibility of the agents, referring to the capability to produce disease or damage, extensive disease or damage, death, infection, or an infection that transmits to others, respectively. Viral biothreats for which EUAs have been issued include SARS-CoV-2, monkeypox virus (mpox), Middle East Respiratory Syndrome Coronavirus (MERS), Zika virus, Ebola virus, H7N9 influenza virus, enterovirus D68, and H1N1 influenza viruses (FDA 2022a, 2023b). *Bacillus anthracis*, the causative agent of anthrax, is a bacterial biothreat for which EUAs have been issued. Non-living biothreats consist of toxins, which are poisonous chemical or biological substances originating from a biological source (Calder and Bland 2015). The need for MCMs against toxin threats is based on a risk assessment of the toxicity, latency, and persistence of the agents, reflecting the amount of the agent needed to be poisonous, the amount of time before toxic effects occur, and how long the agent continues to exist. Examples of toxin biothreats include the anthrax toxin produced by *B. anthracis* and botulinum neurotoxins produced by spore-forming species of the *Clostridium* genus, such as *Clostridium botulinum*. Toxins may also be produced by plants, such as ricin, which naturally occurs in the beans of the castor bean plant, *Ricinus communis*.

Bioterrorism agents and the diseases caused by them are categorized by the CDC based on the threat posed to public health (CDC 2018). Category A agents represent the highest priority agents because they can be easily disseminated or transmitted, produce high mortality rates, have a high potential for major impact on public health, may cause social disruption and panic, and require special public health preparedness efforts. The list of Category A agents/diseases includes *B. anthracis* (anthrax), *C. botulinum* toxin (botulism), *Y. pestis* (plague), variola major (smallpox),

Francisella tularensis (tularemia or rabbit fever), and viruses that cause hemorrhagic fever, including filoviruses such as Ebola and Marburg viruses and arenaviruses such as Lassa and Machupo viruses. Category B agents are moderately easy to disseminate and cause moderate morbidity rates but have low mortality rates, and, while they require the CDC to enhance diagnostic capacity and disease surveillance measures, they do not require the same level of preparedness as Category A agents. Some examples of agents/diseases categorized as Category B include *Burkholderia mallei* and *pseudomallei* (glanders and melioidosis, respectively); alphaviruses that cause encephalitis such as Eastern equine encephalitis (EEE), Venezuelan equine encephalitis (VEE), and Western equine encephalitis (WEE) viruses; and toxins produced by *R. communis* (ricin) and *C. perfringens* (epsilon toxin). Category C agents, the third priority and final category of agents, are emerging pathogens that could be engineered using modern synthetic biology tools to be easily disseminated or transmitted because of the availability of these agents, their ease of production, and their potential to cause high morbidity and mortality rates. Examples of Category C agents include Nipah virus, which can cause encephalitis, and hantavirus, which can cause pulmonary syndrome or hemorrhagic fever with renal syndrome.

HHS has also established and regulates a list of biological select agents and toxins (BSATs) that pose potentially severe threats to the health and safety of the public, some of which overlap with threats to the health and safety of animals and/or plants or their products, which are the responsibility of the United States Department of Agriculture (USDA) (CDC & USDA 2021). The Federal Select Agent Program is responsible for the oversight of the possession, use, and transfer of BSATs through the development, implementation, and enforcement of regulations related to BSATs; maintenance of a national database of entities that handle BSATs; inspection of the facilities of these entities; confirmation that the Federal Bureau of Investigation/Criminal Justice Information Service performs security risk assessments of persons who handle BSATs; and investigation of potential non-compliance incidents involving BSATs. All the Category A agents are currently listed as BSATs, compared with some but not all of the Category B and C agents (CDC & USDA 2022). Some examples of HHS BSATs that are not Category A, B, or C agents or toxins include SARS-associated coronavirus, mpox, tick-borne encephalitis complex (flavi) viruses, tetrodotoxin (a neurotoxin produced by certain marine animals like the fugu pufferfish), and conotoxins (neurotoxins produced by the marine cone snail). Examples of HHS BSATs that overlap with those of USDA include *B. anthracis, B.* mallei and *pseudomallei*, Nipah virus, and VEE. HHS or HHS/USDA Tier 1 BSATs posing the highest risk include all the Category A agents with the exception of Lassa and Machupo viruses, and additionally Variola minor virus, *B. mallei,* and *B. pseudomallei.*

MCM Legislation

Handling and containing dangerous CBRN threats, such as those described in the previous section, is only one of the many challenges inherent to MCM development efforts. Other obstacles include steep costs, technical challenges, programmatic risks, and a typically limited or non-existent commercial market. The federal government may be the only potential customer and funding may be uncertain. A critical unmet need for MCMs to bolster biodefense preparedness was revealed at the turn of the century, which ushered in an era of fear of mass-casualty terrorism in the U.S. (Ackerman and Pate 2001; Bock 2022). The terrorist attacks on September 11, 2001 were followed the next week by the first instances of the anthrax letter attacks via the U.S. mail. Furthermore, intelligence on the interest of terrorist organizations in Category A bioweapons, such as the toxins from *Bacillus anthracis* and *Clostridium botulinum* (the causative agents of anthrax and botulism, respectively), which are among the most potent toxins ever discovered, as well as concerns over the technologies for producing biological and chemical weapons that proliferated from a destabilized Soviet Union (U.S. House of Representatives 2003), galvanized federal efforts to shore up national capabilities to prepare for and respond to PHEs. Legislation has been enacted over the past two decades to incentivize MCM development and, thereby, strengthen national PHE preparedness and response, as described in this section.

The Bioterrorism Act of 2002

The first of these laws in the wake of the September 11 and anthrax letter attacks was the Public Health Security and Bioterrorism Preparedness and Response Act of 2002 (Public Law [P.L.] 107–188), also known as the Bioterrorism Act of 2002. This Act sought to improve national preparedness, response planning, coordination, and reporting for bioterrorism and other PHEs, including a preparedness goal of the development and maintenance of MCMs against biothreats. In addition, this Act provided for the maintenance of a stockpile of drugs, biologics, devices, and other supplies to bolster national biosecurity.

The Project BioShield Act of 2004

In 2004, to further address the unavailability of MCMs for many CBRN threats, Project BioShield was enacted (P.L. 108–276) (Gottron 2009). The Project BioShield Act includes three main provisions: 1) relaxed and expedited procedures for certain CBRN-related spending, 2) guaranteed federal government purchasing of new CBRN MCMs, and 3) emergency use of products that are not approved by the FDA or Department of HHS or that are approved but are indicated for an unapproved use. Project BioShield acquisitions initially included MCM candidates that were at an advanced developmental stage and were directed against agents deemed by the Department of Homeland Security (DHS) to pose a material threat to national security, including anthrax, smallpox, botulinum neurotoxins, and radiological/nuclear agents (Office of Public Health Emergency Preparedness 2006).

Public Readiness and Emergency Preparedness Act of 2005

In December 2005, the Public Readiness and Emergency Preparedness (PREP) Act (P. L. 109–148, Division C, Section 2) was enacted to provide liability immunity to covered persons and to create a compensation program for eligible individuals seriously injured or killed by a covered countermeasure. Through the PREP Act, the

Secretary of HHS is authorized to issue a declaration that provides covered persons liability immunity for claims of losses associated with the use of covered countermeasures for threats or associated diseases or conditions determined to pose a credible risk for a PHE, except in cases of willful misconduct (Office of the ASPR 2021a). Covered persons include those involved with MCM-related activities, such as manufacturers, distributors, and program planners (Department of HHS 2020a). Covered countermeasures include drugs, devices, or biologics that are qualified pandemic or epidemic products, which diagnose, prevent, mitigate, treat, or cure a pandemic or epidemic disease or condition, a serious or life-threatening disease or condition caused by such a product, or a product or technology intended to enhance such a product; security countermeasures determined to be a priority to diagnose, prevent, mitigate, or treat harm from a material CBRN threat or a condition arising from the use of such a product and that the Secretary determines are necessary to protect public health; or any such product issued an EUA. In March 2020, the Secretary issued a PREP Act declaration for immunity from liability for MCMs against COVID-19. PREP Act declarations have also been issued for MCMs against acute radiation syndrome, anthrax, BoNT, nerve agents and certain insecticides, pandemic influenza, and Zika virus (amended January 2023); smallpox, mpox, and other orthopoxviruses (October 2022); Marburg virus (November 2020); and Ebola virus (December 2018) (Office of the ASPR 2023).

All-Hazards Preparedness Act of 2006

In December 2006, the Pandemic and All-Hazards Preparedness Act (PAHPA) (P.L. 109–417) was enacted to provide new authorities for the advanced development and acquisition of MCMs to improve national preparedness and response capabilities for PHEs and national disasters (Office of the ASPR 2019a). One such authority is the Biomedical Advanced Research and Development Authority (BARDA), which supports advanced development and procurement of MCMs to provide a bridge between early development and marketing approval, which has been described as the "valley of death" in pharmaceutical innovation (BARDA 2014). However, BARDA has also supported the development of some FDA-approved products for new uses, such as midazolam injection (Seizalam®; Meridian Medical Technologies LLC), for which the use was expanded to include the treatment of status epilepticus in adults, as may occur upon exposure to a chemical weapon (Laney and Clark 2018). BARDA advances the development of vaccines, therapeutics, and diagnostics for a diversity of CBRN threats, such as botulism, plague, anthrax, smallpox, Ebola virus, and chemical agent exposure. Since 2007, over 40 of the PHEMCE pipeline products have received FDA approval, and over a dozen products have been delivered to the Strategic National Stockpile (SNS) (Office of the ASPR 2019b).

Pandemic and All-Hazards Preparedness Reauthorization Act of 2013

In March 2013, the Pandemic and All-Hazards Preparedness Reauthorization Act (PAHPRA) (P.L. 113-115) reauthorized certain programs under the PHS Act and the FD&C Act regarding all-hazards preparedness, including efforts directed at maintaining and strengthening national preparedness for CBRN and emerging infectious disease threats (Office of the ASPR 2019c; FDA 2018a). This act authorized funding for PHE preparedness programs and funding through 2018 for MCM acquisition under the Project BioShield. PAHPRA also provided flexibility in how state health departments can allocate staff resources in a disaster and increased the flexibility of Project BioShield in supporting MCM advanced research and development. In addition, PAHPRA includes provisions to enhance FDA's authority to support rapid-response efforts for PHEs and enhance MCM review. For example, under PAHPRA, sponsors developing MCMs using the Animal Rule can request two meetings with FDA, including one focused on animal model development and the other occurring before the start of pivotal animal efficacy studies, the protocols for which may be evaluated under special protocol assessment provisions (FDA 2015a). Some of the major program areas addressed by PAHPRA include the National Health Security Strategy, which outlines strategies to strengthen national capabilities for PHE and disaster preparedness and response; Modernization of the National Disaster Medical System, a system that supports national medical response capabilities during PHEs or disasters; and the ASPR, whose mission and key priorities include building a regional disease health response system, sustaining public health security capacity, and improving the MCM enterprise (Office of the ASPR 2019c).

21st Century Cures Act of 2016

In December 2016, the 21st Century Cures Act (P.L. 114–255) was enacted to accelerate and create efficiencies in the development of innovative medical products, including MCMs (FDA 2020a). Funding of over $500 million was authorized over nearly a decade to assist the FDA in fulfilling its duties under this law. The Cures Act supports the modernization of clinical trial design and outcome assessment to expedite clinical development. The Cures Act provides for streamlined MCM procurement processes for BARDA. It also authorizes BARDA to enter into an agreement with a non-profit partner in order to support MCM development efforts. To realize this authority, BARDA established the Division of Research, Innovation, and Ventures (DRIVe), which forms public-private partnerships to accelerate the development of innovative technologies that protect against health security threats using non-dilutive and dilutive funding (Department of HHS 2021a, 2021a). DRIVe included the Rapidly Deployable Capabilities team to respond to COVID-19 using pathogen-agnostic technologies that were previously developed under other programs, including the Early Notification to Act, Control, and Treat program, the goal of which was to inform individuals and healthcare providers early of signs of an infection, and the Beyond the Needle Program to develop alternative delivery systems for vaccines. Other examples include the Redirect and Reboot Programs, which seek to repurpose readily available therapies as MCMs for chemical and biological threats, respectively, and the Digital MCMs Program, which seeks to develop digital tools that can be deployed rapidly for diagnosis and health risk assessment. The Cures Act also established the Regenerative Medicine Advanced Therapy (RMAT) program, which provides an accelerated approval pathway for designated products, and the Breakthrough Devices program, which expedites the review of designated medical devices. The Cures Act provided for the FDA to establish a program for

MCM Priority Review vouchers, which may be used by sponsors who receive them or transferred, including by sale, to other sponsors (FDA 2020a). Material threat MCM Priority Review vouchers were issued for remdesivir (October 2020) and the original-monovalent Pfizer-BioNTech and Moderna mRNA COVID-19 vaccines (August 2021 and January 2022, respectively) (FDA 2022d).

Pandemic and All-Hazards Preparedness and Advancing Innovation Act of 2019

In June 2019, the Pandemic and All-Hazards Preparedness and Advancing Innovation Act (PAHPAIA) (P.L. 116-122) was enacted to advance national health security (Office of the ASPR 2019d). The Secretary, ASPR, and Director of the CDC were given enhanced authorities for PHE preparedness and response efforts. PAHPAIA authorized new programs for PHE preparedness, reauthorized funding for preparedness activities, and enhanced authorities for certain programs, such as the Public Health Emergency Preparedness Cooperative Agreement program. It also authorized uses for the Public Health Emergency Fund in a present or future PHE, advance funding for MCM acquisition under the Project BioShield Act, and funding for MCM advanced research and development efforts. Finally, like PAHPRA, PAHPAIA included provisions to enhance FDA's authority to support rapid-response efforts. Among its many program areas are PHEMCE, ASPR, and accelerating medical countermeasure advanced research and development and supporting procurement.

The Coronavirus Aid, Relief, and Economic Security Act

In March 2020, the Coronavirus Aid, Relief, and Economic Security (CARES) Act was enacted to support response efforts and mitigate the economic toll of COVID-19 (FDA 2021b). Specific to FDA, the CARES Act provided new authorities to FDA intended to give the Agency greater visibility into drug supply chains to help identify, prevent, and mitigate drug shortages. These authorities include requirements for manufacturers of certain drugs to notify the FDA of permanent manufacturing discontinuances or interruption that can meaningfully disrupt supply; to create, maintain, and implement risk management plans that identify and assess risks to drug supply; and to report the amounts of drugs manufactured or otherwise processed for commercial distribution.

MCM Regulatory Authorities

FDA issues regulations as required or authorized by acts of Congress, such as the FD&C Act, which authorizes FDA to regulate food (except for certain poultry, meat, and egg products), human and veterinary drugs, devices for human use, cosmetics, and certain other products, and the MCM laws described in this section, among many others (FDA 2018b, 2018c, 2020b). These regulations, also called rules, reflect the Agency's interpretation of these laws and how to implement them, although the public is provided an opportunity to comment on proposed rules published in the

Federal Register under the Administrative Procedure Act (Carey 2021). The final rules published therein are codified in the Code of Federal Regulations (CFR), which contains 50 areas organized into titles, Title 21 of which contains food and drug regulations (FDA 2018d, 2021c). Regulations carry the force and effect of law, unlike Guidance documents, except where regulations or statutes are cited. As FDA-regulated products, MCMs are subject to FDA regulations, such as those for INDs and marketing applications. Unlike other products, however, MCMs may have special authorities for use in emergencies in addition to traditional pathways, as described in this section.

Emergency Use Authorization

The EUA authority allows FDA to assist with strengthening public health protections against CBRN threats during PHEs by facilitating MCM availability, as described during the FDA Small Business Regulatory Education for Industry (REdI) annual conference sessions titled, "Product Quality Considerations for Emergency Use Authorizations" and "CDER's Role in Public Health Emergency Response & Medical Countermeasure Development," presented on July 19–20, 2021. An EUA is essentially a legal mechanism to facilitate access to MCMs during certain emergencies or a significant potential for an emergency. Early interaction with the FDA is encouraged when a sponsor considers submitting an EUA. Such interactions may involve discussing the product, how to submit information to the Agency, and the contents of the formal submission. As described in FDA Guidance for Industry, "Emergency Use Authorization for Vaccines to Prevent COVID-19," March 2022, sponsors considering submission of an EUA request for an investigational COVID-19 vaccine should contact the Agency as early in development as possible. FDA review includes product quality, facility standards, clinical trial conduct, and data integrity. If it is not feasible for the Agency to verify these characteristics, the Agency may decline to process and review the EUA, and the FDA must prioritize among the EUA requests it receives. Therefore, it is crucial to discuss these topics with FDA as soon as possible.

MCMs that have not been approved, licensed, or cleared, or which have been approved, conditionally approved, licensed, or cleared for a different use, can be authorized for emergency use during an actual or potential PHE involving a CBRN agent (e.g., SARS-CoV-2) or disease/condition caused by such an agent (e.g., COVID-19) provided that certain criteria are met (FDA 2017, 2023b). Section 564 of the FD&C Act, as amended by PAHPRA, gives permission to the FDA Commissioner to issue EUAs for unapproved products or products approved for other uses in certain circumstances. To do so, the HHS Secretary must first make an EUA declaration based on a determination of a domestic or military emergency involving an increased risk of attack with CBRN agent(s), or significant potential for such an emergency; a PHE that affects national security or the health and security of U.S. citizens living abroad involving CBRN agent(s) or an attributable disease or condition, or significant potential for such an emergency; or a material threat sufficient to affect national security or the health and security of U.S. citizens living abroad. Then, after consultation with the ASPR and the Directors of the National Institutes of Health (NIH) and the Centers for Disease Control and Prevention (CDC), as feasible, the

Commissioner may authorize the medical product for emergency use as long as other statutory criteria have been met. Other criteria for EUA issuance include that the CBRN agent can cause a serious or life-threatening disease or condition; the medical product "may be effective" in the diagnosis, treatment, or prevention of the disease or condition, or to mitigate a disease or condition caused by an FDA-regulated product indicated for a disease or condition caused by a CBRN agent; the known and potential benefits of the medical product outweigh its known and potential risks; no other adequate, approved alternative is available for the disease or condition; and other criteria are satisfied that the Secretary may prescribe by regulation (FDA 2017, 2018e).

EUAs may be revised, terminated, or revoked based on the evolution of the threat landscape and accumulating knowledge about the risks and benefits of the MCM. In the event the criteria for an EUA are no longer met after issuance, the EUA can be revoked. For example, a few months after issuance for chloroquine phosphate and hydroxychloroquine sulfate, the EUA was revoked due to clinical data suggesting the drugs may not be effective, and due to serious cardiac events and other adverse events reported with their use, their potential benefits were no longer were believed to outweigh their potential risks (FDA 2020c). By contrast, emerging data for remdesivir continued to support its potential efficacy and positive risk/benefit profile, resulting in the reissuance of its EUA several times in 2020, and ultimately approval following review by the Center for Drug Evaluation and Research (CDER) of the New Drug Application (NDA) for remdesivir (FDA 2020d). After its approval, remdesivir was reissued an EUA because the approval did not cover the entire population for which its emergency use was authorized, including certain hospitalized pediatric patients (FDA 2020e).

Use of a product under an EUA does not require Institutional Review Board (IRB) approval or informed consent under 21 CFR part 50, which can expedite product delivery to patients; however, state and local laws may impose these restrictions (FDA 2017a). For example, recipients of COVID-19 vaccines under an EUA were not required to be given written, signed, and witnessed informed consent; rather, FDA recommended manufacturers of these vaccines provide a Fact Sheet for Recipients and Caregivers with information about the vaccine, such as its ingredients, potential benefits and risks, alternatives, and information on reporting, to ensure vaccinees were informed to the extent practical about the EUA (CDC 2021c; FDA 2017). By contrast, under an Investigational New Drug (IND) or Investigational Device Exemption, an investigational product may be used in a clinical study only after review and approval of the study by an IRB and informed consent of study participants in accordance with 21 CFR part 50.

The COVID-19 pandemic has catalyzed the development of an unprecedented number of MCMs that have been issued EUAs. Approximately one year after the national emergency concerning COVID-19 was declared, COVID-19–related EUAs were issued for over a dozen drugs (FDA 2021a). Small molecules like the antiviral remdesivir and the Janus kinase inhibitor baricitinib (Eli Lilly and Co.) were issued EUAs for the treatment of certain hospitalized COVID-19 patients. At first, remdesivir was issued the EUA only for hospitalized patients with severe COVID-19, which was later revised to include all hospitalized COVID-19 patients regardless of disease severity. Biologics also were issued EUAs including vaccines for the prevention of COVID-19 disease – antibodies for the treatment of mild to moderate COVID-19 in certain patients at risk of progressing to severe COVID-19 or hospitalization (e.g., casirivimab and imdevimab [REGEN-COV®; Regeneron Pharmaceuticals, Inc.] and bamlanivimab and etesevimab [Eli Lilly and Co.]), and blood products like COVID-19 convalescent plasma for the treatment of hospitalized patients with COVID-19. Over 400 medical devices have been issued EUAs since the initial emergency declaration, including devices for blood purification, infusion pumps, PPE and PPE decontamination systems, remote or wearable patient monitoring devices, respiratory assist devices, ventilators and accessories, and other medical devices.

During the 2009 H1N1 pandemic, EUAs were issued for three antivirals, including peramivir (BioCryst Pharmaceuticals, Inc.), oseltamivir phosphate (Tamiflu®; Roche), and zanamivir (Relenza®; GlaxoSmithKline), along with 18 *in vitro* diagnostics and one PPE product (disposable N95 respirators) (FDA 2021p). During other PHEs caused by infectious diseases, EUAs were also issued for *in vitro* diagnostics for Zika virus (14 issued from 2016 to 2017; six revoked), Ebola virus (ten issued from 2014 to 2016 and 2018; three revoked), and MERS-CoV (two issued in 2013 and 2015). The first EUA was issued in 2005 for anthrax vaccine adsorbed for the prevention of inhalation anthrax in adults deemed by the DoD to be at heightened risk of exposure based on the significant potential for an anthrax attack on members of the U.S. military (FDA 2005a). In 2011, a doxycycline mass-dispensing EUA and a National Postal Model EUA were issued for post-exposure prophylaxis of inhalational anthrax, the need for which was replaced by the 2016 FDA Doxycycline Emergency Dispensing Order and the CDC Doxycycline Emergency Use instructions (FDA 2022e). An EUA for an MCM against a chemical threat, specifically the 2mg Atropine Auto-injector (Rafa Laboratories, Ltd.) for the treatment of poisoning by nerve agents and certain insecticides, was first issued in 2017 with amendments into 2018 (FDA 2021a, 2023b). These examples demonstrate the diversity of MCMs, whether investigational or approved for other uses, that can be made available through EUAs to timely respond to PHEs involving biological threats.

Expanded Access

Availability of investigational MCMs to COVID-19 patients during the pandemic has been facilitated by Expanded Access Programs, which may also be requested of FDA outside of a PHE. For example, such programs may be established for patients who cannot enroll in a clinical trial, which may be due to ineligibility, lack of a clinical trial enrolling participants, or excessive distance to the site (FDA 2016, FDA 2019a, 2022f). Under Expanded Access, an individual patient may access an investigational drug or device intended for a serious or life-threatening disease if there is no comparable or satisfactory alternative and the probable risk of the product is not greater than that of the disease as assessed by the patient's physician, among other statutory requirements. Furthermore, widespread access to an investigational drug or device intended for a serious or immediately life-threatening disease or condition may be made available to eligible patients under Expanded Access if there are no comparable or satisfactory alternatives (FDA 2018f). For serious diseases, there must be sufficient evidence of safety and effectiveness to support such use of the product; however, for immediately life-threatening diseases, there need only be a reasonable basis to conclude the product may be effective for its intended use and would not pose an unreasonable or significant risk to the patient.

There are three categories of Expanded Access for drugs including biologics – individual patient, intermediate-size population, and widespread use – each of which may be obtained through either a new IND or a protocol submitted as an amendment to an existing IND, in contrast with an EUA, for which there are neither categories nor IND/protocol submissions (FDA 2016, 2018g). In accordance with 21 CFR 312, subpart I, all categories of Expanded Access must fulfill certain requirements for INDs, such as the reporting of adverse events by the investigator (i.e., licensed physician) to the sponsor, who is responsible for submitting safety reports and annual reports; IRB review of the protocol consistent with 21 CFR 56; informed consent of patients consistent with 21 CFR 50; and maintenance and retention of records in a manner consistent with the requirements of 21 CFR 312, which are not required of EUAs under FDA regulations but which may be required by state or local laws (FDA 2017).

Early in the pandemic, COVID-19 convalescent plasma, an unapproved product that contains antibodies from plasma, was made available to individual patients with serious or immediately life-threatening COVID-19 through emergency Expanded Access in March 2020 (FDA 2021d). The Expanded Access Program for COVID-19 convalescent plasma was transitioned to an EUA in August 2020, with physicians recommended to use the EUA pathway unless patients did not qualify (e.g., pediatric or pregnant patients). Similarly, remdesivir was made available through Expanded Access (NCT04323761; NCT04302766) but was eventually issued an EUA (FDA 2022a). Many other products have been made available to COVID-19 patients through Expanded Access, such as ruxolitinib (Jakafi®; Incyte Corporation), a Janus-associated kinase, for COVID-19 patients with cytokine storm (treatment IND/protocol; NCT04355793); eculizumab (Soliris®; Alexion Pharmaceuticals, Inc.), a recombinant humanized monoclonal antibody against complement protein C5, to prevent immune-mediated death in COVID-19 patients (individual patients, treatment IND/protocol; NCT04288713); and infusion treatment using bone marrow mesenchymal stem cell-derived extracellular vesicle product (ExoFlo™; Direct Biologics, LLC), for patients with COVID-19-associated, moderate-to-severe acute respiratory distress syndrome (intermediate-sized Expanded Access; NCT04657458).

Special Immunizations Program

Vaccine MCMs have been made available to healthy individuals at risk of occupational exposure to biothreats through the Special Immunizations Program (SIP), Division of Medicine, USAMRIID, at Fort Detrick, MD (National Research Council 2011). Biodefense research poses risks of exposure to high concentrations of high-risk pathogens, such as through experiments involving aerosol challenge of laboratory animals with lethal doses of select agents. The SIP was initiated to help protect laboratory workers engaged in biodefense research by providing immunization with investigational vaccines if licensed vaccines were unavailable. Investigational vaccines may be available for at-risk laboratory workers through a cooperative agreement between the SIP and qualified external laboratories (CDC & NIH 2020). These vaccines have thus far exhibited acceptable safety profiles and immunogenicity suggestive of efficacy, although a risk assessment should be conducted in considering whether to offer the vaccine to laboratory personnel. Several clinical trials at the SIP have been posted on ClinicalTrials.gov, including Phase 2 safety and immunogenicity studies of the VEE vaccine (NCT03051386), WEE vaccine (NCT02466750), Rift Valley fever vaccine (NCT03609398), tularemia vaccine (NCT03867162), and EEE vaccine (NCT02654509).

Approval or Licensure

Less than a year after the COVID-19 PHE was declared, remdesivir was approved for the treatment of COVID-19 in adults and pediatric patients 12 years and older weighing at least 40 kg requiring hospitalization (FDA 2020d). Remdesivir, an adenosine analogue, as well as chloroquine and hydroxychloroquine, antimalarials, were found to inhibit SARS-CoV-2 *in vitro* shortly after cases were first reported in Wuhan, China (Wang et al. 2020; Liu et al. 2020). Remdesivir, also known as GS-5734, had previously been identified as a potential antiviral against SARS-CoV, MERS-CoV, and bat CoVs (Sheahan et al. 2017) and was being clinically developed as a treatment for Ebola virus disease (Mulangu et al. 2019). In November 2020, remdesivir was approved based in part on the NIAID-sponsored Adaptive COVID-19 Treatment Trial (ACTT-1), a multicenter, adaptive, randomized, double-blind, placebo-controlled trial of the safety and efficacy of remdesivir for the treatment of COVID-19, with a total enrollment of 1062 participants (Beigel et al. 2020). This trial demonstrated that hospitalized COVID-19 patients administered intravenous remdesivir (200-mg loading dose on Day 1, with 100-mg maintenance doses on Days 2–10 or until discharge or death) recovered significantly faster than participants administered placebo, with median days to recovery of 11 days compared with 15 days, respectively. Furthermore, remdesivir exhibited an acceptable safety profile in hospitalized COVID-19 patients. FDA approval of remdesivir indicates that the drug was determined by the Agency to be safe and effective for its intended use and its benefits outweigh its risks when used as intended.

Approved/licensed therapeutic MCMs are available for a limited number of BSATs, including Ebola virus (FDA 2021e), botulism, anthrax, plague, smallpox, and tularemia (FDA 2019b). For example, two antitoxins are available to treat botulism. In 2013, botulism antitoxin heptavalent (A, B, C, D, E, F, G) – (equine) (BAT®) was approved for the treatment of adults and pediatric patients with symptomatic botulism after documented or suspected exposure to BoNT/A, B, C, D, E, or F, the seven known BoNT serotypes (Emergent BioSolutions Canada Inc. 2020). BAT consists of a mixture of immune globulin fragments from the pooled plasma of horses immunized with a botulinum toxin or toxoid (i.e., an inactivated toxin that remains immunogenic). In 2016, botulism immune globulin intravenous (human) (BabyBIG®) was approved for the treatment of infant botulism caused by toxin types A or B in infants (California Department of Public Health 2021). BabyBIG comprises anti-botulinum neurotoxin (BoNT) antibodies derived from the pooled plasma of human donors who were immunized with an investigational botulism vaccine and produced high titers of neutralizing antibodies against the BoNT/A and /B serotypes. Novel botulism MCMs are being developed that can facilitate scale-up and surge capacity in the event of a BoNT-related emergency or that otherwise offer potential benefits. These include MCMs such as recombinant human or humanized monoclonal antibodies, which block further BoNT binding to neuronal cells and facilitate their clearance; small-molecule or peptide-based inhibitors of BoNTs, which can act intracellularly to antagonize BoNT activity; as well as treatments that improve neuromuscular function (Patel et al. 2018;, Rasetti-Escargueil and Popoff 2019; Vazquez-Cintron et al. 2020; Snow et al. 2021).

Several approved/licensed therapies are available for the treatment of inhalational anthrax in combination with appropriate antibacterial drugs, including anthrax immune globulin intravenous (human) and two monoclonal antibodies for injection (FDA 2019c). Similar to the botulism immune globulin products, anthrax immune globulin intravenous (Anthrasil®) is a polyclonal immune globulin antitoxin that was initially approved in 2015 (Emergent Manufacturing Operations Baltimore LLC 2018). Raxibacumab is a human IgG1λ antibody that was initially approved in 2012 (Emergent Manufacturing Operations Baltimore LLC 2021), and obiltoxaximab (Anthim®) is a chimeric IgG1κ antibody that was initially approved in 2016 (Elusys Therapeutics Inc. 2019). Rather than exerting antibacterial activity directly, these antibodies in these products bind to the protective antigen of *B. anthracis* and prevent it from binding to its receptors, thereby preventing intracellular entry of lethal and edema factors, which are the enzymatic components that produce toxicity. Both raxibacumab and obiltoxaximab are also approved for prophylaxis of inhalational anthrax when alternative therapies are either unavailable or inappropriate. Antibiotics approved for post-exposure prophylaxis for documented or suspected exposure to *B. anthracis* spores include ciprofloxacin, doxycycline, levofloxacin, and penicillin G procaine (FDA 2019d). Doxycycline and ciprofloxacin FDA Emergency Dispensing Orders were issued in 2016 to permit emergency dispensing of approved oral dosage forms of these antibiotics without prescriptions in the event of a PHE involving *B. anthracis* (FDA 2021a).

Various antibiotics have been approved for plague and tularemia. Antibiotics indicated for both diseases include doxycycline and other tetracyclines and streptomycin, although the latter is not readily available in the U.S. and causes serious toxicities (FDA 2019b; Apangu et al. 2017). Additional antibiotics have been approved for plague, including the fluoroquinolones ciprofloxacin, levofloxacin, and moxifloxacin. First-line treatments for plague include the fluoroquinolones, which are bactericidal like streptomycin, as well as off-label use of gentamicin, which is an effective alternative to streptomycin (CDC 2020b). Doxycycline is only bacteriostatic and has neither concentration-dependent activity nor post-antibiotic effects, so its efficacy may be limited in serious plague cases. The first and only treatment for smallpox was approved in 2018. This oral antiviral therapy, tecovirimat monohydrate capsule (TPOXX®), is a small-molecule inhibitor of the orthopoxvirus VP37 envelope (SIGA Technologies, Inc. 2022).

To date, vaccines have been licensed for fewer than half of the HHS and HHS/USDA Tier 1 BSATs, including Ebola, anthrax, smallpox, and plague, although the plague vaccine is not currently available because it was discontinued by the sole manufacturer (FDA 2019e). A vaccine for the prevention of Ebola (Ervebo®) was approved by the FDA in 2019 (Merck Sharp & Dohme LLC 2022). Five pandemic H1N1 influenza vaccines received FDA approval (FDA 2021e). Before the H1N1 pandemic, the threat of highly pathogenic avian H5N1 influenza resulted in development efforts leading to the approval of the first H5N1 influenza vaccine (Sanofi Pasteur 2007) in 2007 for inclusion in the Strategic National Stockpile, with other such vaccines following, such as one approved in 2013 (ID Biomedical Corporation of Quebec 2013) for commercial distribution and another, AUDENZ™, in 2020 (Seqirus Inc. 2020) for distribution to the U.S. Government as part of national pandemic preparedness (FDA 2021e). FDA support for the development of new vaccines and other MCMs against CBRN and infectious disease threats is covered in the next section.

FDA REGULATION OF MCMS

FDA Regulatory MCM Mission Space

FDA support for the development, availability, and accessibility of MCMs is aligned with its mission to protect public health, including ensuring the safety, efficacy, and security of human medicines, biological products, and medical devices (FDA 2018h). Specifically, FDA evaluates the safety and efficacy of MCMs for civilian and military populations; helps to safeguard MCM availability during PHEs by monitoring the supply chain for potential disruption and potentially adulterated or counterfeit products; and, in collaboration with the CDC, monitors the safety and efficacy of MCMs during and after their deployment (FDA 2018h, 2021f). Other response activities, such as those during the COVID-19 pandemic, may relate to the global supply chain, fraudulent drug products, and public communications, as discussed in the FDA REdI Annual Conference 2021 session titled, "CDER's Role in Public Health Emergency Response & Medical Countermeasure Development," presented on July 20, 2021. For instance, FDA has issued over 100 warning letters to internet pharmacies and manufacturers of products with drug claims regarding fraudulent COVID-19 products. FDA can also take the enforcement action of seizure or injunction of health fraud. Another example of an FDA response activity during the COVID-19 pandemic includes urging consumers not to use certain products, such as certain adulterated hand sanitizer products, which are regulated as over-the-counter drugs (FDA 2021g). Furthermore, the Agency alerted certain pharmaceutical manufacturers and pharmacists during the COVID-19 pandemic about the potential public health hazard of ethanol or isopropyl alcohol that is contaminated or substituted with methanol, which caused cases of fatal poisoning following ingestion of contaminated alcohol-based hand sanitizers, as described in FDA Guidance for Industry, "Policy for Testing of Alcohol (Ethanol) and Isopropyl Alcohol for Methanol, Including During the Public Health Emergency (COVID-19)," January

2021. The Agency also warned consumers about hand sanitizer packaged to appear as food or drink, as this could lead to accidental ingestion and potentially death, especially for children (FDA 2021g). In addition to these actions, FDA has reviewed hundreds of thousands of submissions during the COVID-19 pandemic. CDER alone received ~342,000 in Fiscal Year (FY) 2020, and as of June 20, 2021, CDER had received ~249,000 submissions, as described in the FDA REdI Annual Conference 2021 session titled, "Electronic Submissions Update," presented on July 19, 2021.

Other examples of the ways FDA supports the development, availability, and accessibility of MCMs can be found in the Agency's accomplishments reported for FY 2019–2020, some of which included implementing the Material Threat MCM Priority Review Voucher program; addressing barriers to the availability of unapproved MCMs for CBRN threats through EUAs and expanded access; supporting development of national capabilities to collect and analyze MCM-related data in PHEs to enable real-time decision-making about their safety and effectiveness; and authorizing use of MCMs past their expiration date to prevent critical drug shortages (FDA 2021f), which is possible through the issuance of an EUA, exercise of enforcement discretion, extension of the expiration date of eligible approved MCMs that have been stockpiled under PAHPRA, or through the Shelf-Life Extension Program, a federal program administered by the DoD in which stockpiled drugs for military or contingency use (e.g., antidotes for nerve agents, antivirals, and antibiotics) are stability tested by FDA (FDA 2020f).

FDA works with local, state, national, and international governmental bodies and academic, non-profit, and industry organizations engaged in MCM-related research, development, and lifecycle management (FDA 2020g). For example, FDA's Coronavirus Treatment Acceleration Program (CTAP), an emergency program to expedite the development of COVID-19 therapies, includes an email inbox for developers to route their inquiries to the appropriate point of contact, promoting early and frequent interactions with the Agency. In addition, CTAP provides subject matter expertise on regulatory standards for drugs and clinical trial design to the Accelerating COVID-19 Therapeutic Interventions and Vaccines (ACTIV) government and industry partnership. FDA is a core member of PHEMCE, whose mission is to advance the nation's capabilities to prepare for and respond to CBRN threats, whether intentional or accidental, and emerging infectious diseases by coordinating medical countermeasure-related efforts within HHS and in cooperation with PHEMCE interagency partners.

FDA conducts activities to better support codified PHE preparedness and response efforts through its Medical Countermeasures Initiative (MCMi) (FDA 2019f). The MCMi program is managed by the Office of Counterterrorism and Emerging Threats (OCET). FDA launched the MCMi program in 2010 to promote MCM development through the establishment of clear regulatory pathways for MCMs, policies and mechanisms to expedite access to available MCMs, and tools to assist with regulatory decisions related to MCMs. In FY 2019, FDA obligated over \$100 million to MCM-related activities, including base funding of approximately \$25 million for the MCMi program, \$34 million for CBRN preparedness, \$44 million for pandemic influenza preparedness, and \$5 million in no-year funding for emerging health threat preparedness and response efforts. Through MCMi, FDA has established Public Health and Security Action Teams to resolve technical and regulatory issues in developing MCMs, such as the Microbial Sequencing and Multiplex *In Vitro* Diagnostics Action Team, FDA/DoD Enhanced Engagement Action Team, and Acute Radiation Syndrome Action Team (FDA 2019f). MCMi Regulatory Science Program activities have recently included developing organ-on-chip models for evaluating therapeutic MCMs for acute radiation syndrome, methods to obtain data from patients administered MCMs during PHEs, methods to decontaminate respiratory protective devices so they can be reused, improved animal models for emerging infectious diseases, tools to assess the efficacy of Ebola virus MCMs, and a biobank of samples from Ebola patients and vaccinees to characterize immune correlates of protection against Ebola virus (FDA 2021f).

FDA supports the development, availability, and accessibility of MCMs through such activities as providing regulatory and technical advice to MCM industry and government partners developing MCMs; reviewing applications for MCMs and providing approval, licensure, or clearance if standards are met; supporting access to unapproved MCMs through another mechanism, as appropriate; maintaining the MCMi Professional Development program to help Agency staff maintain critical skills to support MCM activities; and maintaining the MCMi Regulatory Science Program, which is responsible for developing tools, standards, and approaches to evaluate MCMs (FDA 2019f). For example, in FY 2019, FDA approved most of the MCM applications that were under review, such as the Biologics License Applications (BLAs) for smallpox and mpox vaccine, live, non-replicating (Jynneos®; Bavarian Nordic A/S) and dengue tetravalent vaccine, live (Dengvaxia®; Sanofi Pasteur, Inc.); the NDA for baloxavir marboxil (Xofluza®; Genentech, Inc.) for the treatment of uncomplicated influenza; BLA supplements to extend the indications of Afluria® and Afluria® Quadrivalent influenza vaccines (Seqirus Pty Ltd.) and to include a new presentation for Fluzone® Quadrivalent influenza vaccine (Sanofi Pasteur, Inc.); and numerous influenza tests. FDA also issued an EUA for a rapid test for Ebola virus detection and granted or acknowledged 18 amendments for EUA diagnostic devices.

FDA works directly with MCM sponsors and applicants through formal and informal meetings, in which FDA answers questions and provides advice on product development, such as chemistry, manufacturing, and controls; nonclinical and clinical development activities; data analysis; and regulatory matters, described in greater detail in the next section (FDA 2021h). In FY 2021, FDA held 127 formal meetings with MCM sponsors/applicants of Prescription Drug User Fee Act (PDUFA) products, including 37 at the Center for Biologics Evaluation and Research (CBER) and 18 at CDER, with 215 and 15 non-PDUFA meetings at the centers, respectively (representing a substantial increase from 47 and 15 such meetings at CBER and CDER, respectively, in FY 2019 (FDA 2019f)); 72 formal meetings at the Center for Devices and Radiological Health; and additional pre-EUA and EUA meetings. FDA has also issued dozens of draft and final Guidance documents addressing MCM-related issues, which indirectly supports MCM developers by providing the Agency's current thinking on specific topics, as discussed in the section entitled "FDA Guidance Documents Relevant to MCMs." Other support for developers includes hosting

Advisory Committee meetings, workshops, training courses, and symposia, which are available to the public. For example, in FY 2019, FDA hosted the Vaccines and Related Biological Products Advisory Committee, the purpose of which was to discuss and recommend strains for an influenza virus vaccine for the upcoming influenza season in the southern hemisphere. Background materials are made publicly available before or after these Advisory Committee meetings, and interested members of the public are permitted to present information, ideas, and opinions about the issues the committee is discussing either in writing or by a brief formal oral presentation (FDA 2019f). FDA provides live webcasts of the Advisory Committee meetings free of charge. Public workshops provide a forum for input from a wide range of stakeholders, including scientific and technical subject matter experts, regulators, and the general public. Some examples of public workshops and training courses hosted in FY 2019 related to MCMs include the Cutaneous Radiation Injuries workshop, Filovirus Animal Nonclinical Group workshop, Achieving Data Quality and Integrity in Clinical Trials Involving High-Consequence Pathogens training course, and Achieving Data Quality and Integrity in Maximum Containment Laboratories training course. The FDA Science Forum public symposium offered an opportunity to discuss research at FDA, including research related to MCMs, such as rapid-response strategies for infectious disease outbreaks, the Animal Rule pathway, and rapid diagnostic tests. In FY 2021, there was a shift to COVID-19-related meetings, such as the Vaccines and Related Biological Products Advisory Committee to discuss COVID-19 vaccines and webinar series on PPE and COVID-19 tests (FDA 2021h).

FDA Meetings with MCM Developers

CBER, which regulates certain biological products (e.g., gene therapies, vaccines, allergenic extracts, antitoxins/antivenins, blood and blood products, cellular therapies) among other products (FDA 2018i)), consults with prospective innovators/product developers early in development. For example, the CBER Advanced Technologies Team (CATT) fosters early product development efforts by responding to inquiries related to substantially novel advanced manufacturing technologies that pose unique regulatory or technical hurdles but may offer improvements in the manufacturing, characterization, quality, safety, and/or efficacy of CBER-regulated products (FDA 2019g). CATT meetings should focus on innovative technologies that can significantly impact product development, manufacturing, or controls. CBER also awards grants to support research on improving the manufacturing process and controls for biologics. The CDER Small Business and Industry Assistance lends support to small pharmaceutical business and industry developers of drugs and certain biologics regulated by CDER (including monoclonal antibodies for *in vivo* use; therapeutic proteins, except for vaccines and blood products; immunomodulators; and growth factors, cytokines, and monoclonal antibodies intended to alter cell production *in vivo*) (FDA 2022g, 2022h). Specifically, FDA assists developers in navigating and understanding the information provided by FDA on drug product development, regulation, and review by responding to inquiries, offering webinars, hosting conferences and workshops, providing web-based tutorials, emailing educational updates, and publishing an electronic newsletter.

Upon request, PDUFA meetings are granted to sponsors of MCMs or other products barring a compelling reason to deny the request, such as development not being at the appropriate stage for such a meeting to be productive or insufficient information in the background package to support meaningful discussion (FDA 2019h). Meetings can assume a face-to-face, tele/videoconference, or written responses only (WRO) format and are categorized as Type A, B, C, D, and INTERACT (CBER 2022). Type A meetings include those required to address serious safety concerns or stalled product development, such as a clinical hold. Type B meetings include pre-IND meetings, pre-EUA meetings, and pre-NDA/BLA meetings, among others. Type B End-of-Phase meetings include specific End-of-Phase meetings and End-of-Phase 2/Pre-Phase 3 meetings. Type C meetings are any type other than A, B, B (EOP), D, or INTERACT regarding product development and review. Type D meetings focus on no more than two topics, involving no more than three disciplines or Divisions, such as a narrow issue or a general question about a novel development approach with only a few questions or a follow-up question raising a new issue following a formal meeting. INTERACT meetings address novel product-specific (or product-derivation strategy focused) challenges prior to a Pre-IND (Type B) meeting and are meant to facilitate IND-enabling activities. Requests for Type A, Type B End-of-Phase, and Type D meetings/WROs have the fastest response timeline at 14 days, and these meetings/WROs are scheduled within 30, 70, or 50 days of request, respectively. Requests for other Type B, Type C, and INTERACT meetings/WROs have a response timeline of 21 days, with the meetings/WROs scheduled within 60 days for Type B or 75 days for Type C or INTERACT. Meeting packages for Type A, Type C for early consultation on a new surrogate endpoint, Type D, and INTERACT meetings must be submitted with the meeting request, whereas Type B, Type B End-of-Phase, and Type C briefing packages must be submitted within 30, 50, or 47 days, respectively, of the scheduled meeting/WRO.

Once product development has progressed to the Pre-IND stage, the FDA will provide formal early consultation to sponsors of drug products intended to treat illnesses that may be life-threatening or severely debilitating, in accordance with 21 CFR Sec. 312.82(a). Outside scientific experts or advisory committee members may also be invited. The main objective of the Pre-IND meeting is to provide an opportunity for FDA-reviewing officials to review and to reach an agreement upon the design of IND-enabling studies, such as nonclinical toxicology studies. Prior to the meeting, the sponsor submits questions and a briefing package on the drug product's development, including chemistry, manufacturing, and controls (CMC) information; any *in vitro* or *in vivo* efficacy data or safety data; proposed plans for the Phase 1 clinical study; and any plans for pediatric studies (FDA 2019h). Sponsors may also inquire about the optimal presentation and formatting of the data and other information in the IND for FDA concurrence. Following IND submission, in accordance with 21 CFR 312.40–42, the FDA reviews the data and other information to ensure that the proposed Phase 1 clinical study does not pose an unreasonable or significant risk to the safety of human subjects, the clinical investigators are sufficiently qualified by experience and training and informed by the investigator's brochure to conduct the study, and other statutory requirements are met. During this review period, FDA may communicate with the sponsor about any deficiencies in the IND, request clarification, or advise as to the adequacy of the technical data and clinical plans upon request for advice or comment.

Once Phase 1 clinical data are available, in accordance with 21 CFR 312.82(b), FDA officials will review the data and meet with sponsors to agree on the design of Phase 2 clinical trials to ensure that they will provide data that are sufficient for the Agency to make a determination on the product's safety and effectiveness in the future marketing application. This End-of-Phase 1 meeting also offers sponsors the opportunity to discuss pediatric studies. In accordance with 21 CFR 312.47(b)(1), agreements reached at the meeting on the objectives and design of clinical studies, the adequacy of technical information to support the studies and/or marketing application, and the design and timing of pediatric studies will be recorded in minutes taken by the Agency and provided to the product sponsor. These minutes and other documents provided to the sponsor permanently record any agreements reached in the meeting, and FDA will presume the objectives and design of studies conducted in accordance with the agreement are sufficient to obtain marketing approval provided there are no significant scientific developments that would require otherwise. Therefore, these meetings are indispensable risk mitigation tools for sponsors of investigational new drugs to gain the Agency's concurrence on the adequacy of technical information and clinical plans.

The procedures outlined in Section 312.47(b)(1) are used for both End-of-Phase 1 and End-of-Phase 2 meetings. In accordance with 21 CFR 312.47(b), the FDA will also meet with sponsors in an End-of-Phase 2 meeting or Pre-NDA/BLA meeting. These meetings help to expedite development by providing sponsors with binding advice on what information and data are necessary to support the marketing application, thereby avoiding non-value-adding activities, as well as how to best present the information to facilitate Agency review. End-of-Phase 2 meetings are primarily intended for sponsors of new molecular entities or approved drugs and biologics with new uses, although any IND sponsor may request this type of meeting. Items for discussion can include plans and protocols for Phase 3 studies, additional nonclinical studies, pediatric assessments, tentative labeling, or any issue relevant to proceeding to Phase 3 or the adequacy of the existing data to support the future marketing application. Pre-NDA/BLA meetings provide a forum for discussion of any unresolved issues, such as deficiencies that could result in a refuse-to-file action; the adequate and well-controlled studies upon which the sponsor intends to rely for approval; the status of pediatric assessments, if any; methods for data analysis and presentation; and formatting of the application. Before these meetings, the sponsor must timely submit a background package summarizing the history and current status of product development, plans for additional studies with synopses or protocols, and information to support waivers or deferrals of pediatric assessments, if applicable. The objectives and design of studies that are conducted according to agreements reached in Pre-NDA/BLA meetings are considered sufficient for the purpose of obtaining

FDA approval provided there are no significant scientific developments that would alter the Agency's assessment.

FDA Guidance Documents Relevant to MCMs

FDA Guidance documents, once finalized, reflect FDA's current thinking on how requirements of statutes and regulations for specific topics can be satisfied (FDA 2022i). The recommendations in Guidance documents are non-binding on the Agency and the public, and alternative approaches may be pursued so long as they satisfy all applicable requirements. However, Guidance for Industry documents are meant to support development efforts, and following the recommendations can expedite and constrain costs in product development through avoidance of studies or other activities that may not satisfy regulatory requirements. Typically, draft Guidance documents that are not for implementation undergo a period of public comment after the announcement of the draft in the Federal Register, after which the draft is revised or finalized; however, FDA's Good Guidance Practices indicate that public participation will only be held before Guidance implementation if feasible and appropriate, in accordance with Section 701(h)(1)(C) of the FD&C Act (21 U.S.C. 371(h)(1)(C)) and 21 CFR 10.115(g)(2).

Approximately one year after the national emergency concerning COVID-19 was declared, the FDA issued 60 Guidance documents for industry, FDA staff, and other stakeholders on human drugs, biologics, and medical devices that were related to COVID-19. Early in the PHE, the FDA posted the final Guidance for industry and FDA staff, "Enforcement Policy for Ventilators and Accessories and Other Respiratory Devices During the Coronavirus Disease 2019 (COVID-19) Public Health Emergency," March 2020, followed soon thereafter by Guidance documents such as "Policy for Certain REMS Requirements During the COVID-19 Public Health Emergency Guidance for Industry and Health Care Professionals," March 2020; "Product-Specific Guidances for Chloroquine Phosphate and Hydroxychloroquine Sulfate," April 2020; and "Exemption and Exclusion from Certain Requirements of the Drug Supply Chain Security Act During the COVID-19 Public Health Emergency," April 2020 (FDA 2021j).

Numerous FDA Guidance documents on regulatory pathways may be highly relevant to the development of COVID-19 or other MCMs, depending on the product and indication, such as the following final Guidance for Industry documents: "Expedited Programs for Serious Conditions – Drugs and Biologics," May 2014; "Product Development Under the Animal Rule," November 2015; "Expanded Access to Investigational Drugs for Treatment Use – Questions and Answers," June 2016"; "Emergency Use Authorization of Medical Products and Related Authorities," January 2017; and "Breakthrough Devices Program," December 2018 (FDA 2020h). The draft Guidance for industry, "Material Threat Medical Countermeasure Priority Review Vouchers," January 2018, may also be of considerable interest to MCM developers, as it answers questions the Agency has received about the Material Threat MCM priority review voucher program, such as eligibility criteria and requirements for award. FDA also issues Guidance documents on specific types of products that may be used as MCMs, such as the following final Guidance for Industry documents: "Internal Radioactive Contamination – Development of Decorporation Agents," March 2006; "Influenza: Developing Drugs for Treatment and/or Prophylaxis," April 2011; "Antibacterial Therapies for Patients With an Unmet Medical Need for the Treatment of Serious Bacterial Diseases," August 2017; "Anthrax: Developing Drugs for Prophylaxis of Inhalational Anthrax," May 2018; "Extending Expiration Dates of Doxycycline Tablets in Strategic Stockpiles," April 2019; "Smallpox (Variola Virus) Infection: Developing Drugs for Treatment or Prevention," November 2019; and "Limited Population Pathway for Antibacterial and Antifungal Drugs," August 2020.

Several FDA Guidance documents pertain to orphan drug and humanitarian use regulations and designations, which are often relevant to MCM development (FDA 2018j). For example, there are final Guidance documents available on the following topics: "Interpreting Sameness of Monoclonal Antibody Products Under the Orphan Drug Regulations," April 2014; "Clarification of Orphan Designation of Drugs and Biologics for Pediatric Subpopulations of Common Diseases," July 2018; "Rare Diseases: Common Issues in Drug Development," February 2019; "Humanitarian Use Device (HUD) Designations," September 2019; and "Interpreting Sameness of Gene Therapy Products Under the Orphan Drug Regulations," September 2021. Several draft Guidance documents are also available in this area, including "Rare Diseases: Early Drug Development and the Role of Pre-IND Meetings," October 2018, and "Rare Diseases: Natural History Studies for Drug Development," March 2019.

Potential Regulatory Designations and Pathways for MCMs

Orphan Designation

The Orphan Drug Act of 1983 was created in recognition of the need to financially incentivize the development of drugs and biologics for rare diseases or conditions, as development costs may exceed profits given the paucity of customers (FDA 2018k). While the designation does not change the requirement for approval, benefits include tax credits up to 25% for qualified clinical studies; seven years of market exclusivity; waiver of the prescription drug user fees, which are over $3 million if clinical data are required, unless other indications that are not Orphan Designations are included; among other incentives (FDA 2022j). In addition, pediatric assessments are not required for marketing applications for drugs with an Orphan Designation, as long as they are not also seeking a non-Orphan-Designated indication (FDA 2005b). In accordance with the Orphan Drug regulations in 21 CFR Part 316, the disease or condition for which the drug is intended must affect fewer than 200,000 people in the U.S., or the drug must be administered to fewer than 200,000 people annually for vaccines, preventative drugs, or diagnostic drugs. There may be an "Orphan Subset" of a patient population that otherwise numbers 200,000 or more, in which case it must be demonstrated that these other patients are not appropriate candidates. Otherwise, the criterion is a reasonable expectation that the research and development costs of the drug for the specific indication cannot be recovered in the U.S. market. The sponsor must request Orphan Designation in accordance with 21 CFR 316.20.

While not all MCMs meet these criteria, such as the COVID-19 vaccines, the diseases or conditions associated with CBRN threats and emerging infectious diseases are often rare. For example, approximately 100–200 cases of botulism are reported annually in the U.S. (CDC 2019). Accordingly, both BAT and BABYBIG have received Orphan Designation for the treatment of botulism and infant botulism, respectively, and both these products are approved for the designation (FDA 1989, 2011). There are also products developed for CBRN threats that have received Orphan Designation but are not FDA approved, such as the recombinant botulinum vaccine A/B for prevention of botulism caused by inhalational intoxication with BoNT/A1/B1 (FDA 2019i). FDA anticipates that some drugs for CBRN threats will meet the criteria for Orphan Designation, including drugs that are eligible for a Material Threat MCM Priority Review Voucher (FDA 2018l). For example, TPOXX, which was the first product to be awarded the Material Threat MCM Priority Review Voucher, also received Orphan Designation (FDA 2018o).

Humanitarian Use Designation

The Safe Medical Devices Act of 1990 included a provision to create a regulatory pathway for devices intended for rare diseases or conditions, with the goal of addressing the challenge inherent in generating sufficient clinical evidence to demonstrate efficacy in patient populations that are small by definition (FDA 2022k). This pathway, the Humanitarian Device Exemption program, exempts pre-market approval applications for HUDs from the effectiveness requirements of Sections 514 and 515 of the FD&C Act. The FDA review period is 75 days with a Humanitarian Device Exemption, as compared with 180 days for a pre-market approval application without a panel or 320 days with a panel (FDA 2022l). In accordance with 21 CFR 814.3(n), an HUD is a "medical device intended to benefit patients in the treatment or diagnosis of a disease or condition that affects or is manifested in not more than 8,000 individuals in the United States per year." Rather than a reasonable assurance of effectiveness, the criterion is the demonstration of probable benefit (FDA 2022l). In addition, rather than needing to provide a reasonable assurance of safety, the criterion is the lack of exposure to an unreasonable, significant risk of harm by the device. However, the labeling must reflect that the effectiveness of the device has not been demonstrated, and the device can only be used at facilities with IRB oversight. Furthermore, there are restrictions on selling an HUD for profit following a Humanitarian Device Exemption approval, unless certain criteria are met (e.g., a pediatric indication). While device MCMs could be eligible for HUD, this designation has been sought infrequently for these products.

Expedited Programs

Expedited programs are available to expedite the availability of drugs (including biologics) intended to address an unmet medical need for serious or life-threatening diseases or conditions, including Accelerated Approval, Breakthrough Therapy Designation, Fast Track Designation, Priority Review, and RMAT Designation

(FDA 2014). In addition, the Breakthrough Device Designation is meant to incentivize the development of innovative devices for life-threatening or irreversibly debilitating diseases or conditions (FDA 2022m). A serious disease or condition is one that is associated with morbidity that is persistent, recurrent, or irreversible and that substantially impacts day-to-day functioning or survival or that will become serious if left untreated (FDA 2014). An unmet medical need may be an immediate or long-term need for a defined population or society in general that is not adequately addressed by available products. The lack of an available drug for a serious disease or condition constitutes an unmet medical need, although a drug may qualify based on other benefits, such as improved safety, efficacy, patient compliance, and/or availability compared with available drugs, defined as approved or licensed drugs relevant to the U.S. standard of care for the indication. The regulations in 21 CFR part 312, Subpart E, were the first to address how to expedite the availability of drugs for serious diseases and conditions for which there are no satisfactory alternatives. MCMs often fulfill the criteria for one or more of these programs and may receive more than one designation. For example, remdesivir received both Fast Track and Priority Review designations (FDA 2020l), as did TPOXX (FDA 2018o). Summaries of the expedited programs are provided in this section.

Accelerated Approval

The Accelerated Approval Program is provided for by Section 506(c) of the FD&C Act, as amended by Section 901 of the Food and Drug Administration Safety and Innovation Act (FDASIA) of 2012 (FDA 2014). This program enhances FDA's ability to promote innovation to expedite the availability of safe and effective products. Over 250 drugs and biologics have been approved through the Accelerated Approval pathway, many of which are oncology and human immunodeficiency virus drugs, although some are MCMs (FDA 2021k). For example, ciprofloxacin (CIPRO®; Bayer HealthCare Pharmaceuticals, Inc.) was approved through this pathway for the treatment of inhalational anthrax post-exposure (FDA 2022n). To qualify for Accelerated Approval, the drug must be indicated for a serious condition, as in the other three expedited programs, as well as provide a meaningful benefit over available drugs (FDA 2014). An additional requirement, which is also an advantage of this program, is that the drug must demonstrate an effect on a surrogate endpoint or intermediate clinical endpoint. A surrogate endpoint is a measure that is reasonably likely to predict, but does not itself measure, clinical benefit, while an intermediate clinical endpoint is one that can be measured sooner than irreversible morbidity or mortality and that is also reasonably likely to predict clinical benefit. For example, serum ciprofloxacin levels were the surrogate endpoint for ciprofloxacin, with supportive clinical evidence obtained during the anthrax letter attacks (Bayer HealthCare Pharmaceuticals, Inc. 2021). FDA considers the disease process and the relationship of the drug's effect to it in evaluating the predictive potential of these endpoints (FDA 2014). A determination on the reasonable likelihood of clinical benefit is made by FDA on a case-by-case basis.

Approval based on a surrogate or intermediate clinical endpoint is advantageous because fewer participants may be necessary to demonstrate the effect than would be required for a clinical endpoint, i.e., an endpoint related to how a patient feels, functions, or survives (FDA 2014). In addition to being smaller, these studies may be shorter or fewer. This advantage is more pronounced if there is a long period before the clinical benefit would be evident, such as with long-term diseases, or if a large clinical trial would be necessary to demonstrate benefit, such as with rare diseases. The appropriateness of this program should be discussed with the Agency early in development. If the product meets the criteria for traditional approval, Accelerated Approval will not be granted. The same standards for safety and effectiveness that are required for traditional approval also apply to Accelerated Approval in accordance with Section 505(d) of the FD&C Act. Conditions of approval under this program include requirements to submit promotional materials to FDA during preapproval review and to conduct postmarketing confirmatory trials with due diligence, which should be underway when the marketing application is submitted if use of this pathway is clear during development. Approval may be withdrawn if a confirmatory trial does not verify clinical benefit, the applicant fails to conduct the confirmatory trial with due diligence, or evidence suggests the product is not effective under the conditions of use.

Breakthrough Therapy Designation

The Breakthrough Therapy Designation is provided for by Section 506(a) of the FD&C Act, to which this designation was added by Section 902 of FDASIA (FDA 2014). As for the other expedited programs, to qualify for Breakthrough Therapy Designation, the drug must be intended for a serious or life-threatening disease or condition. Furthermore, preliminary clinical data must support that the drug may offer an improvement over available drugs on a clinically significant endpoint that measures an effect on serious disease symptoms or irreversible morbidity and mortality, although a well-substantiated pharmacodynamic biomarker may be considered. The expectation is that the preliminary evidence will be from Phase 1 or Phase 2 trials; however, nonclinical studies could also support clinical evidence of pharmacologic activity. Preliminary clinical evidence based on comparison to an available drug would be ideal, although historical data that are well documented may also be acceptable. As for all drugs, approval is still based on a review of full data and statutory standards for establishing safety and efficacy for the intended use. Substantial evidence is a matter of judgment considering the magnitude of the effect as well as its clinical importance compared with an available therapy or placebo, or historical data if there is no available therapy.

The Breakthrough Therapy Designation request can be submitted with the IND or afterward; however, to capitalize on the benefits of the designation, the request should be sent by the End-of-Phase 2 meeting (FDA 2014). These benefits include intensive Agency support, including timely guidance and interactive communications on how to make drug development efforts more efficient (e.g., advice on study design and data analysis) and an organizational commitment including senior management to ensure a collaborative and proactive review is conducted, with the intention for a cross-disciplinary staff member to lead the review and maximize its efficiency. In addition, Breakthrough Therapy Designation affords the opportunity for applicants to submit portions of their application for rolling review by the Agency before the entire application has been submitted. Nevertheless, if at any point emerging data suggest the criteria for the designation are no longer met, it may be rescinded by FDA or withdrawn by the sponsor.

Breakthrough Device Designation

The Breakthrough Devices Program is provided for by Section 515B of the FD&C Act, which was created by Section 3051 of the Cures Act and amended by Section 901 of the FDA Reauthorization Act of 2017 (P.L. 115-52) (FDA 2018m). This program has similarities with the Breakthrough Therapy Designation in both eligibility criteria and benefits. Eligible products include devices or device-led combination products that are more effective for life-threatening or irreversibly debilitating diseases or conditions that satisfy at least one of the following criteria: breakthrough technology; provision of an option where no approved or cleared option exists; significant advantages over existing approved or cleared options; or the availability of the device is in the best interest of patients (FDA 2021l). This program replaced the Priority Review and Expedited Access Pathway programs for medical devices, with any device granted this designation considered to be a Breakthrough Device by the FDA. Sponsors of Breakthrough Device-designated products can discuss device and data development plans with the FDA through program options, such as sprint discussions, and they can also request a clinical protocol agreement and receive prioritized review. An example of a product granted Breakthrough Device designation during the COVID-19 pandemic is a neurostimulation device, the pdSTIM™ System (Stimdia Medical, Inc.), designed to wean patients off mechanical ventilatory support (Business Wire 2021).

Regenerative Medicine Advanced Therapy Designation

The Cures Act also established the RMAT program, which provides an accelerated approval pathway for RMAT products (FDA 2020l). Under Section 3033 of the Cures Act, products eligible for RMAT designation include regenerative medicine therapies, such as cell therapies and human cell and tissue products, that demonstrate the potential to meet an unmet medical need for a serious or life-threatening disease or condition based on preliminary evidence from clinical studies (FDA 2023c). FDA interprets Section 506(g) of the FD&C Act, which was added by Section 3033 of the Cures Act, to mean that human gene therapies and xenogeneic cell products may constitute RMATs. The request for this designation should be submitted with the original IND or as an IND amendment. Sponsors of RMAT-designated products can discuss possible surrogate or intermediate endpoints that can accelerate approval with FDA early in product development. To date, over 80 products have received RMAT designation (FDA 2021m, 2023d).

Fast Track Designation

The Fast Track designation program is provided for by Section 50-6(b) of the FD&C Act, which was added by the Food and Drug Administration Modernization Act of 1997 and amended by Section 901 of FDASIA (FDA 2014). To qualify for Fast Track designation, the drug must be intended for a serious or life-threatening disease or condition, as with the other expedited programs. In addition, nonclinical or clinical data must demonstrate that an unmet medical need can potentially be addressed by the drug. Otherwise, it must be designated as a qualified infectious disease product under the Generating Antibiotic Incentives Now Act, which was passed as part of FDASIA to incentivize the development of new antibacterial and antifungal drugs to address antibacterial resistance (Department of HHS 2017). Such drugs must either treat a serious or life-threatening infection caused by a treatment-resistant pathogen or be a qualifying pathogen under Section 505E(f) of the FD&C Act. A list of qualifying pathogens is provided in 21 CFR 317.2. The request for this designation can be submitted with the IND or afterward, but to capitalize on the benefits of the designation, the request should not be submitted after the pre-NDA or BLA meeting (FDA 2014). These benefits include frequent meetings with the FDA review team and potential eligibility for Priority Review. During the COVID-19 pandemic, several COVID-19 vaccine candidates received Fast Track designation, including those developed by Pfizer/BioNTech (Pfizer Inc. 2020), Moderna (Forbes 2020), and Novavax (Novavax, Inc. 2020). Various other drug candidates also received this designation, such as ZYESAMI® (aviptadil), a synthetic form of human vasoactive intestinal polypeptide that is a potential treatment for acute lung injury/acute respiratory distress syndrome in COVID-19 patients (NRx Pharmaceuticals, Inc. 2023), along with Brilacidin, a small molecule host defense peptide/protein mimetic that has received Fast Track designation not only as a potential treatment for COVID-19 but also for oral mucositis and, under the Qualified Infectious Disease Product designation, as a potential treatment for bacterial skin infections (Innovation Pharmaceuticals, Inc. 2021).

Priority Review

Priority Review was originally provided for in 1992 by PDUFA, which was reauthorized through FY 2027 by the Food and Drug Administration Reauthorization Act of 2012 (FDA 2018n, 2023e). To qualify for Priority Review, the original application or efficacy supplement must be for a drug that is intended for a serious or life-threatening condition, as with the other expedited programs (FDA 2014). In addition, the application or efficacy supplement must be for a drug that constitutes a significant improvement in safety or efficacy or that has been designated a qualified infectious disease product. Otherwise, it must be a supplement that proposes a change to the labeling based on a pediatric study report under section 505A of the FD&C Act or an application or supplement submitted with a Priority Review Voucher, such as a Tropical Disease, Rare Pediatric Disease, or Material Threat MCM Priority Review voucher. The FDA assigns the designation at filing regardless of whether it is requested, although it may be requested by the applicant. The benefit of the Priority Review program is that the goal date for FDA to take action on Priority Review applications is six months instead of the standard ten-month review period, which may result in faster entry to market. Furthermore, Priority Review vouchers are transferable and may be sold for tens of millions or over a hundred million dollars based not only on expedited review but also on the value from capturing competitors' market share and lengthening the time from approval to launch of generic competitors (Ridley and Régnier 2016).

The Material Threat MCM Priority Review Voucher program was provided for by Section 3086 of the Cures Act, which added Section 565A of the FD&C Act (FDA 2018l, 2020i). The goal of this program is to incentivize the development of new pharmaceutical MCMs for material threats that are eligible for priority review. Material threats are those listed in the annual PHEMCE Strategic Implementation Plan that have been determined by the Secretary of the DHS to materially threaten national security. To be eligible for this program, the product must be intended for the prevention or treatment of harm from a CBRN agent, or an MCM for a CBRN agent, that has been determined by the DHS to be a material threat; eligible for Priority Review; approved after the Cures Act was enacted; and constitute a drug or combination product with an active ingredient not previously FDA approved. Remdesivir was awarded a Material Threat MCM Priority Review Voucher in 2020 (FDA 2020l), as have several other MCMs, such as tecovirimat (TPOXX; SIGA Technologies, Inc.), an antiviral for the treatment of smallpox, in 2018; smallpox and mpox vaccine (Jynneos®; Bavarian Nordic A/S), a live non-replicating vaccine for the prevention of smallpox and mpox, in 2019 (FDA 2018o); and atoltivimab, maftivimab, and odesivimab-ebgn (Inmazeb™; Regeneron Pharmaceuticals, Inc.), a combination of human monoclonal antibodies for the treatment of Ebola virus infection, in 2020 (FDA 2020a, 2020j,). The fee rate for using an MCM Priority Review Voucher in Fiscal Year 2023 was approximately $1.5 million (FDA 2022d). FDA's draft Guidance for Industry, "Material Threat Medical Countermeasure Priority Review Vouchers," January 2018, addresses the program (FDA 2018l).

Animal Rule

The Animal Rule regulations are codified in 21 CFR 314.600-650 (drugs) and 21 CFR 601.90-95 (biologics), which went into effect in July 2002 (FDA 2015a). Approval of MCMs for diseases caused by CBRN threats often requires efficacy studies to be conducted in animals. The Animal Rule regulations allow for establishment of the efficacy of new drugs or biologics for the amelioration or prevention of serious or life-threatening conditions in adequate and well-controlled animal studies when clinical trials are unethical and field studies are infeasible. Safety must still be evaluated in human participants under the pre-existing requirements. For example, clinical efficacy studies involving challenges with smallpox would not be ethical because of the potentially fatal nature of the disease, and field studies are not feasible because the disease has been eradicated by the smallpox vaccine. Over a dozen CDER drug products have been approved under the Animal Rule, including products for biological threats, such as ciprofloxacin (CIPRO; Bayer HealthCare Pharmaceuticals, Inc.), levofloxacin (LEVAQUIN; Janssen Pharmaceuticals, Inc. 2018), and moxifloxacin (AVELOX; Bayer Healthcare Pharmaceuticals, Inc.) for plague; obiltoxaximab (ANTHIM; Elusys Therapeutics, Inc.) and raxibacumab (Human Genome Sciences, Inc.) for anthrax; and tecovirimat monohydrate (TPOXX; SIGA Technologies, Inc.) as well as brincidofovir (TEMBEXA®; Chimerix, Inc.) (FDA 2021n) for smallpox. CBER products approved under the Animal Rule include BAT for botulism, anthrax immune globulin intravenous (human) (Anthrasil; Cangene) for inhalational anthrax in combination with appropriate antibacterial drugs, and anthrax vaccine adsorbed (BIOTHRAX; Emergent BioSolutions) for post-exposure prophylaxis following suspected or confirmed *B. anthracis* exposure when administered in combination with recommended antibacterial drugs. The anthrax vaccine is the only vaccine to date approved under the Animal Rule regulations.

Under the Animal Rule, substantial evidence of effectiveness can be generated from animal studies provided four criteria are met: 1) the pathophysiologic mechanism by which the agent or toxin produces toxicity is reasonably well understood; 2) efficacy is demonstrated in more than one animal model expected to react to the agent or toxin in a way that is predictive of the response in humans, or only one such model if it is well characterized; 3) the endpoint in the animal studies is clearly related to the desired human benefit (e.g., improved survival); and 4) relevant data from animals and humans allows for the selection of an effective dose in humans (FDA 2015a). For example, the efficacy of the anthrax vaccine as post-exposure prophylaxis for anthrax was established under 21 CFR Part 601, Subpart H, by demonstrating that the vaccine induced toxin-neutralizing antibody levels in humans estimated to confer a 70% survival probability, as extrapolated from protective toxin-neutralizing levels in rabbits and non-human primates challenged with a lethal dose of aerosolized anthrax spores (FDA 2015b). Thus, a species-independent assay for an immunologic marker can serve as a "bridge" between the animal efficacy studies and human immunogenicity studies. To support approval, Animal Rule products must also satisfy the following requirements: 1) A plan or approach to conduct post-marketing studies must be included in the licensing application to evaluate the safety and efficacy of the product if such studies become ethical and feasible; 2) the labeling must clearly indicate that efficacy studies are entirely based on animal studies along with other relevant information, such as directions for use, reasonably foreseeable risks, and adverse reactions; and 3) restrictions may be placed on the product to ensure its safe use, such as a requirement for health care practitioners to have special training or for patients to be followed up in a specific manner (FDA 2015a).

The FDA encourages early and frequent communications with the Agency, public health officials, and the military, as applicable, due to the complexity inherent in the development of Animal Rule products (FDA 2015a). FDA may request input from Advisory Committees on Animal Rule-related issues such as the acceptability of the animal model and the design of the animal efficacy studies intended to support approval. Sponsors of Animal Rule MCMs can meet with FDA to discuss and agree upon the size and design of the adequate and well-controlled animal efficacy studies in a process called a Special Protocol Assessment under Section 505(b)(5)(B) of the FDA&C Act, as amended by PAHPRA. Such sponsors may also receive two Type C Animal Rule meetings with FDA, one to discuss issues related to developing animal models and one before starting the pivotal animal studies under Section 565(d) of the FD&C Act (21 U.S.C. 360bbb-4(d)), as amended by PAHPRA. Animal Rule products may also be eligible for Fast Track and Priority Review or Orphan Drug Designation; however, they are not eligible for Breakthrough Therapy Designation because this designation statutorily requires clinical evidence under Section 506 of the FD&C Act (21 U.S.C. 356). Animal efficacy data can support EUA or Expanded Access, which the FDA decides on a case-by-case basis considering the emergency situation, population size, risk-benefit analysis, and whether the data support the use of these mechanisms.

THE PUBLIC HEALTH EMERGENCY MEDICAL COUNTERMEASURES ENTERPRISE

Overview of PHEMCE Mission

The lifecycle of MCMs involves basic research, advanced development and manufacturing, procurement, inventory management, stockpiling, deployment, distribution, dispensing, administration, monitoring, evaluation, and assessment, throughout which regulatory science management activities occur (Office of the ASPR 2015). PHEMCE was created in 2006 by HHS to coordinate the breadth of MCM-related efforts required to effectively prepare for and respond to a national emergency (Office of the ASPR 2017). These efforts extend from basic research to lifecycle management and require intensive collaboration within and between federal agencies; state, local, tribal, and territorial governments; and industry, non-profit, and academic organizations (Office of the ASPR 2015). The Commissioner of the FDA is one of three core HHS members of the PHEMCE. The other two core members are the Directors of the CDC and the National Institute of Allergy and Infectious Diseases (NIAID) of the NIH. Led by the HHS Office of the ASPR, PHEMCE is a governance structure that coordinates MCM-related efforts within HHS and across senior leadership from various federal interagency partners, such as the DoD and the DHS (Department of HHS 2019). This section traverses the lifecycle of MCMs and the roles played by PHEMCE partners, with examples provided from the COVID-19 pandemic as well as other threats.

Requirements Setting

The ASPR leads efforts in setting requirements for civilian MCMs (Department of HHS 2023a). Requirement-setting is guided by threat and risk assessments conducted by the DHS and medical consequence and public health response assessments based on the input of all PHEMCE partners and subject-matter experts. The ASPR Office of Policy and Planning manages the process for requirements setting using expert input, including information related to these risk assessments. This information has included scenario-based analyses, such as policy analyses and modeling results, which provide information about which product categories are needed and the numbers of individuals who would use these products during a PHE; an integrated capabilities document, which describes the necessary capabilities of the health care system and any gaps along with the quantities of MCMs expected to be effectively used in a PHE; product-specific requirements, including the target characteristics of the products that are required and the goals for stockpiling; and research requirements, which detail the target characteristics of MCMs needed for threat areas of unmet need to inform research (Office of the ASPR 2014). These requirements have guided resource allocation for the spectrum of activities required to develop MCMs and acquire them by PHEMCE partners.

Early-Stage Research

MCM-focused research and development efforts include early and advanced-stage efforts to discover lead candidates, develop manufacturing strategies and controls for them, and produce them in sufficient quantities (Department of HHS 2023b). The NIH is the PHEMCE lead for the conduct and funding of basic research on threat agents, the diseases or conditions they cause, and MCMs to combat them (Department of HHS 2023c). In the COVID-19 pandemic, NIH led the HHS research response on COVID-19 through early-stage *in vitro*, *in vivo*, and epidemiologic research on the virus and disease as well as clinical trials of MCMs, such as vaccines, monoclonal antibodies, and small molecules, among others (NIAID 2020). While there are hundreds of potential COVID-19 MCMs, there are insufficient resources to test all of them in adequately powered clinical trials in an operationally relevant timeframe (NIH 2021). To meet an unmet need for a collaborative, efficient forum to prioritize MCM candidates, expedite their clinical testing, improve the capacity and coordination of clinical trials, and accelerate evaluation of candidates for authorization or approval, the NIH launched the Accelerating COVID-19 Therapeutic Interventions and Vaccines (ACTIV) partnership in collaboration with the Foundation for the NIH. ACTIV includes federal partners such as the FDA, CDC, BARDA, and DoD, international partners such as the European Medicines Agency, and private industry partners with the goal of developing a coordinated COVID-19 research response.

In addition to the U.S. Department of Agriculture, the DoD is a non-HHS federal partner in PHEMCE that also supports MCM research and development (Department of HHS 2023d). For example, the Defense Advanced Research Projects Agency (DARPA) is a DoD agency that supports research and development on potentially breakthrough technologies for national security (DARPA 2023a). For example, DARPA developed a Pandemic Prevention Platform (P3) that was launched in 2017 to rapidly discover, characterize, produce, test, and deliver nucleic acid-based MCMs directed against infectious disease targets, with performers including both biopharmaceutical companies (AbCellera and AstraZeneca) and academic institutions (Duke and Vanderbilt Universities) (DARPA 2020). The P3 platform can deliver these MCMs at the breakthrough pace of fewer than 60 days from threat identification, in contrast with the years required for traditional processes (DARPA 2023b). The FDA authorized an antibody discovered by AbCellera and NIAID in the P3 program, bamlanivimab (LY-CoV555 [Eli Lilly]), for emergency use in the treatment of certain patients with mild-to-moderate COVID-19 at risk of progression to severe disease based on data from the BLAZE-1 (NCT04427501) trial (FDA 2021r). Other DoD agencies also support MCM research, such as the Defense Threat Reduction Agency (DTRA), which supports research on weapons of mass destruction, the threat environment, developing cutting-edge MCMs, and counter-threat capabilities

(DTRA 2023). For example, in collaboration with the Defense Innovation Unit, DTRA developed a technology called Rapid Analysis of Threat Exposure (RATE; Phillips) that warns of SARS-CoV-2 infection 48 hours before the onset of symptoms (Philips 2020; Vergun 2020). A non-invasive, wearable device collects data on 165 vital signs and other biomarkers and analyzes them using the RATE COVID-19 algorithm, which was developed using artificial intelligence and machine learning on an over 40,000-case dataset.

Advanced Development and Manufacturing

Once lead candidate MCMs are identified, they move to the advanced development and manufacturing stage. BARDA promotes advanced research and development of MCMs against CBRN threats, pandemic influenza, and emerging infectious diseases, including but not limited to support for manufacturing optimization, validation, and scale-up (BARDA 2023).

BARDA had over 70 MCMs in its COVID-19 pipeline in the first quarter of 2021, including diagnostics (e.g., assays to detect SARS-CoV-2 antigens or antibodies), therapeutics (e.g., anti-SARS-CoV-2 antibodies, COVID-19 convalescent plasma, and hyperimmune globulin), vaccines (e.g., mRNA, viral vector, and subunit vaccines), rapidly deployable capabilities (e.g., biomarkers, algorithms, platforms, and devices), and other technologies (e.g., a single-administration microneedle skin patch for pandemic vaccines from the University of Connecticut) (Department of HHS 2021b). The DoD also supports advanced MCM development and manufacturing efforts. For example, the DoD's Joint Program Executive Office for Chemical, Biological, Radiological and Nuclear Defense facilitates advanced development of MCMs for Warfighters (JPEO-CBRND 2023). Total COVID-19 support and assisted acquisition exceeded 75 billion USD obligated across over 200 contracts, including vaccines, diagnostics, therapeutics, PPE, enablers (e.g., syringes, swabs), industrial base expansion, and international donations (JPEO-CBRND 2022).

Regulatory Science Management and Monitoring, Evaluation, and Assessment

Throughout the research and development stage, the FDA works with prospective MCM developers and sponsors through formal and informal meetings, workshops, training, and other communications, which are described in greater depth in the section entitled "FDA Regulation of MCMs." The FDA's primary roles within PHEMCE are to ensure that MCMs are safe and effective before they are used and to lead regulatory science management, including evaluating regulatory aspects of MCM development and utilization (Department of HSS 2023). FDA has approved over 350 MCMs since 2012, thereby bolstering national preparedness for anticipated threats (FDA 2022o). However, in a PHE with a novel threat, approval timelines are likely unacceptably slow, whereas the issuance of EUAs can expedite the availability of novel MCMs outside an investigational protocol. In FY 2021 alone, the FDA issued 180 EUAs that enabled access to over 575 COVID-19-related medical products (FDA 2021h). The FDA continues reviewing evidence of the safety and effectiveness of MCMs during and after their deployment (FDA 2022p). For example, the FDA revokes EUAs if scientific evidence indicates that the known and potential risks are not outweighed by the known and potential benefits. During the COVID-19 pandemic, EUAs for certain *in vitro* diagnostics were revoked when these devices exhibited evidence of inaccuracy, and the EUAs for hydroxychloroquine sulfate and chloroquine phosphate were revoked four months after they were issued based on accumulating data suggesting they were ineffective for COVID-19 and caused serious side effects, such as cardiac arrhythmias (FDA 2021p).

Procurement, Inventory Management, and Stockpiling

MCMs may be needed for threats that are intentionally released (e.g., a terrorist attack or intentional dumping of CBRN agents) or accidentally released due to human error, technical failure, or natural disaster (e.g., chemical spill/leak from a facility or vehicle or inadvertent release of a pathogen) (National Library of Medicine 2020). Unlike a hazmat incident, which typically involves accidental release of a chemical that is not a widespread disaster, a CBRN incident is more likely to be a deliberate act intended to cause mass casualties. Therefore, MCMs need to be available in large quantities to support national preparedness and response capabilities. Section 564B of the FD&C Act, as amended by PAHPRA, permits the U.S. Government to pre-position MCMs in anticipation of FDA approval, licensure, or clearance; EUA issuance; or investigational use authorization so that they can be deployed rapidly in an emergency situation (FDA 2017a). To maintain the stockpile, MCMs must not only be acquired but also stored, handled, and transported appropriately under the correct conditions to ensure their continued quality (Office of the ASPR 2021b). In order to manage the potentially massive and fast-moving quantities of MCMs required in a PHE, the SNS developed an Inventory Management and Tracking System to help public health agencies track MCMs deployed from the SNS.

The SNS is the national repository of medical supplies, such as airway management equipment and ventilators, and pharmaceuticals, such as antibiotics, vaccines, and antitoxins, which is overseen by the ASPR (Department of HHS 2021c). CDC leads the procurement of MCMs and the management of their inventory, including stockpiling and replenishing supplies (Department of HHS 2022). PHEMCE has stockpiled over a dozen MCMs in the SNS, which can furnish MCMs within 12 hours (Office of the ASPR 2020). The MCMs from the SNS can supplement state and local repositories on a short-term basis in an emergency until adequate quantities are available (Office of the ASPR 2021c). For example, the stockpiled antimalarials chloroquine/hydroxychloroquine were to be distributed from the SNS following their authorization for emergency use for COVID-19 (FDA 2020k). The SNS has deployed PPE products, such as N95 respirators, face shields, and disposable gloves and gowns across the U.S. and territories, as well as ventilators to high-need areas (Office of the ASPR 2021d). In addition to the SNS, CHEMPACKs also contain national MCM inventory at pre-positioned locations across the U.S. for faster response in the event of a chemical attack. An EUA may be required for products in the SNS or CHEMPACKs. For example, in 2017, the FDA issued an EUA for emergency use of the Rafa Laboratories, Ltd. Atropine Auto-injector as an initial anti-muscarinic treatment for poisoning by nerve agents or certain insecticides (FDA 2017b). The EUA was issued following a request by the CDC due to a lack of FDA-approved products available to replenish the inventory upon their impending expiration, and the product was approved the following year (Rafa Laboratories, Ltd. 2019).

Response Planning, Policy Guidance, and Communication

Development of federal operational and response plans, utilization policies and guidance, and communication strategies is coordinated by the CDC and ASPR (Department of HHS 2022). CDC and ASPR are also responsible for coordination of the development of plans for federal response and strategies regarding allocation of MCMs. In recent years, CDC has led efforts to integrate social media into mechanisms to share PHE messages and to improve the capacity to develop easily understood and accessible materials to aid in message comprehension by the public (CDC 2021d). CDC works with community partners to identify challenges and opportunities in risk response communication to enhance efficiencies and response capacity.

Distribution and Administration

The COVID-19 pandemic has revealed the challenges inherent in rapidly distributing and administering mass quantities of MCMs to the public. These activities are jointly led by the ASPR and CDC, which also develop policies and guidance for MCM prioritization, allocation, and use (Department of HHS 2023a). Meticulous coordination has been required between federal, state, local, tribal, and territorial government bodies and vaccination sites, including private healthcare organizations, public health clinics, pharmacy networks, community centers, correctional facilities, and religious institutions, in addition to health insurers and many other stakeholders, commensurate with the scale and complexity of the PHE (CDC 2020c). The initial plan in a PHE may follow an all-hazards approach but later account for the unique features of the MCM, such as its specific supply constraints and storage and handling conditions, the scalability of the distribution and administration plans, and communication of prioritization schemes and logistics to healthcare providers and the public (Department of HHS 2021d). For example, MCMs that are thermostable, require only a single dose, and can be administered by a convenient and simple route may be prioritized over those requiring ultra-cold storage, multiple doses, or an invasive route of administration. While supplies are limited, MCM delivery is tightly focused on high-priority risk groups, which typically receive the intervention in a closed setting, such as delivery of COVID-19 vaccines first to healthcare providers and residents of long-term care facilities (Department of HHS 2020b). As supply becomes less constrained, the MCM delivery is expanded to lower-priority groups, and once there is excess supply, MCMs are made universally available through a broad network to enhance accessibility.

Distributing MCMs requires four steps: authorization, prioritization, allocation, and distribution (CDC 2021e). Following authorization by the FDA, a prioritization scheme is recommended by the CDC for which groups should receive the MCM in what order. For example, during the COVID-19 pandemic, vaccines were initially in short supply. Therefore, based on the recommendation of the Advisory Committee on Immunization Practices, an independent expert panel, the CDC recommended that the first groups to be allocated COVID-19 vaccines should be healthcare personnel and residents of long-term care facilities (Phase 1a); followed by frontline essential workers and people 75 years of age and older (Phase 1b); people 65 to 74 years of age, people 16 to 64 years of age with underlying medical conditions, and other essential workers (Phase 1c); and lastly, the rest of the population. This scheme was based on the goals of decreasing morbidity and mortality, preserving social functioning, and reducing the excess burden of COVID-19 on individuals experiencing disparities. However, individual jurisdictions ultimately decide how to distribute the vaccines, which may be different than as recommended by the CDC. Specific quantities of COVID-19 vaccine doses are then allocated to all 50 U.S. states, territories, and other jurisdictions based on the number of adults living there proportionate to the entire U.S. population (CDC 2021f). Then, the jurisdictions decide where to distribute the doses, such as health departments, hospitals, and pharmacies.

The SNS is critically involved with efforts to distribute and administer MCMs. A year after the national emergency was declared, the SNS had deployed over 4,000 trucks and 600 flights carrying thousands of tons of cargo in response to the COVID-19 pandemic (Office of the ASPR 2021d). In partnership with the McKesson Corporation, the SNS produced and distributed ancillary and mixing kits containing needles, syringes, alcohol prep pads, and other supplies needed for vaccines that require reconstitution in support of Operation Warp Speed (Department of HHS 2020b, Department of HHS 2020c). For the Pfizer-BioNTech COVID-19 vaccine, dry ice kits with dry ice gloves, scoops, safety cards, and other supplies were also distributed to support shipment at ultra-cold temperatures. To help with treating allergic reactions that can occur with vaccination, the SNS distributed 100,000 epinephrine autoinjectors to vaccination clinics outside hospital settings across the U.S. and territories (Office of the ASPR 2021e). Federal medical stations are managed by the SNS to support efforts to administer MCMs (Office of the ASPR 2021d). These caches include beds and a three-day supply of medicines, medical devices, and other supplies that can be rapidly deployed to transform a building into a shelter for 50–250 primary or critical care patients and to support quarantine efforts.

CONCLUSION

Summary

While biological agents of natural origin or from biological weapons development have long been recognized as a significant potential threat to the safety of the public, legislative efforts to address this concern began to accelerate in the early 2000s. Several factors drove this effort, including the anthrax attacks in 2001, general concerns over terrorism from both hostile regimes and non-state entities, and concerns over biosecurity following the collapse of the Soviet Union, a state known to have performed bioweapons research. Additionally, natural events, such as the bird flu scare, highlighted the need for a robust ability to respond to biological agents. Given that commercial markets for MCMs are often small and uncertain, the U.S. Government recognized the urgent need to fund development, manufacturing, procurement, and ongoing storage of MCMs in the SNS. Additionally, the FDA developed regulations to enable evaluation and usage of MCMs via non-traditional pathways to allow for their availability against novel threats in a timeframe relevant for a PHE.

The Bioterrorism Act of 2002 and the Project BioShield Act of 2004 were important first steps that sought to increase the nation's responsiveness to public health emergencies by funding a stockpile of MCMs against known threats, increasing funding, and decreasing regulatory burdens for MCM development. The All-Hazards Preparedness Act of 2006 expanded on this mission through the creation of BARDA to support the advanced development and procurement of MCMs. In 2006, PHEMCE was created to coordinate MCM efforts with both the CDC and NIAID of the NIH. PHEMCE covers the lifecycle of MCMs, beginning with the funding of basic early-stage research through the NIH via NIAID and the DoD via organizations such as DARPA and DTRA. Potential candidate MCMs then proceed to Advanced Development and Manufacturing, funded by agencies such as BARDA and the DoD's JPEO-CBRND. Further action can include procurement and stockpiling of MCMs in advance of FDA approval to ensure an adequate supply in the SNS. Finally, PHEMCE can coordinate the logistical distribution of MCMs.

In addition to legislative efforts to support MCM development and supply, the FDA also began the MCMi managed by OCET. A key component of this effort has been to support MCM developers by providing new avenues for timely interaction and feedback from the Agency. In addition, a variety of regulatory pathways exist that enable MCMs to be utilized significantly faster than conventional approval and licensure timelines. A variety of pathways exist because some tools, such as Orphan Drug status or utilization of the Animal Rule, may not apply in all situations. Other Expedited Programs potentially applicable to MCMs include Breakthrough Therapy, Accelerated Approval, and Fast-Track Designations. In addition, the FDA has special authorities applicable specifically to MCMs. These include EUA, Expanded Access, and the SIP.

Outlook

The COVID-19 pandemic demonstrated the clear need for continued investment in preparedness against biological threats. The response to the pandemic highlighted several successes derived from the creation of agencies and regulatory pathways developed over the last two decades. For example, NIAID-funded basic research on platform RNA vaccine technology led to a rapidly developed vaccine, which was then able to be manufactured in large quantities and procured via BARDA-funded efforts. BARDA additionally funded the development of monoclonal antibodies that were able to receive EUA from the FDA and be used in the treatment of COVID-19 within six to nine months. These are just two examples among many that could be highlighted.

Given the devastating impact of the pandemic in both human and economic terms, it is safe to assume that future legislation will ensure expanded funding to support rapid MCM development. Obvious targets for future investment will likely include expanded basic research into novel platform technologies that enable a rapid development cycle and expanded manufacturing capacity capable of being quickly activated. Additional areas for future investment highlighted by the pandemic response might focus on ensuring that the supply chain required to support surge manufacturing capacity is sufficient. High-profile shortages in syringes and vials demonstrated this need, but supply chain issues plagued all aspects of the MCM production process and are having a lingering impact on current manufacturing for both emergency response and ongoing drug production. Additional challenges requiring future investment involve ensuring that sufficient logistical support for MCM cold storage and distribution is in place to support future emergency response.

REFERENCES

Ackerman, G., and J. Pate. 2001. *Assessing the Threat of Mass-Casualty Bioterrorism*. James Martin Center for Nonproliferation Studies at the Middlebury Institute of International Studies at Monterey. October 1. https://www.nti.org/analysis/articles/mass-casualty-bioterrorism/.

Apangu, T., et al. 2017. "Successful Treatment of Human Plague with Oral Ciprofloxacin." *Emerging Infectious Diseases* 23 (3): 553–555.

Ball, P. 2021. "The Lightning-fast Quest for COVID Vaccines — and What it Means for Other Diseases." *Nature* 589: 16–18.

BARDA. 2014. *Project BioShield Annual Report*. Washington, DC: U.S. Department of HHS, Office of the ASPR.

BARDA. 2023. *Core Services*. January. https://medicalcountermeasures.gov/barda/core-services/.

Bayer HealthCare Pharmaceuticals, Inc. 2021. "Prescribing Information for CIPRO® (ciprofloxacin)." November. https://www.accessdata.fda.gov/spl/data/e1614736-4aa8-42d3-9fd2-bce239a04364/e1614736-4aa8-42d3-9fd2-bce239a04364.xml#ID_2484df6e-c995-4c8e-ba5d-5cf49f717a33.

Beigel, J. H., et al. 2020. "Remdesivir for the Treatment of Covid-19 - Final Report." *New England Journal of Medicine* 383 (19): 1813–1826.

Bock, Eric. 2022. "2001 Anthrax Attacks Revealed Need to Develop Countermeasures Against Biological Threats." *NIH Record* (NIH) LXXIV (10). https://nihrecord.nih.gov/2022/05/13/2001-anthrax-attacks-revealed-need-develop-countermeasures-against-biological-threats.

Business Wire. 2021. CORRECTING and REPLACING Stimdia Medical Receives FDA Breakthrough Device Designation for the pdSTMT(TM) System Designed to Facilitate Weaning Patients from Mechanical Ventilation: https://www.businesswire.com/news/home/20210318005897/en/CORRECTING-and-REPLACING-Stimdia-Medical-Receives-FDA-Breakthrough-Device-Designation-for-the-pdSTIM%E2%84%A2-System-Designed-to-Facilitate-Weaning-Patients-from-Mechanical-Ventilation

Calder, A., and S. Bland. 2015. "Chemical, Biological, Radiological and Nuclear Considerations in a Major Incident." *Surgery (Oxford)* 33 (9): 442–448. doi:10.1016/j.mpsur.2015.07.006.

California Deptartment of Public Health. 2021. "Prescribing Information for BabyBIG [human botulinum neurotoxin a/b immune globulin]." June. https://www.fda.gov/media/74647/download.

Carey, M. P. 2021. *An Overview of Federal Regulations and the Rulemaking Process.* Washington, DC: Congressional Research Service. https://fas.org/sgp/crs/misc/IF10003.pdf.

CBER. 2022. "SOPP 8101.1: Regulatory Meetings with Sponsors and Applicants for Drugsand Biological Products." October 1. https://www.fda.gov/media/84040/download#:~:text=Type%20D%20Meeting%20%2D%20A%20meeting,than%203%20disciplines%20or%20Divisions.

CDC & NIH. 2020. *Biosafety in Microbiological and Biomedical Laboratories.* Washington, DC: Department of HHS. https://www.cdc.gov/labs/pdf/SF__19_308133-A_BMBL6_00-BOOK-WEB-final-3.pdf.

CDC. 2018. *Bioterrorism Agents/Diseases.* April 4. https://emergency.cdc.gov/agent/agentlist-category.asp.

CDC. 2019. *National Botulism Surveillance.* June 6. https://www.cdc.gov/botulism/surveillance.html.

CDC. 2020a. "Medical Countermeasure Readiness." *CDC Website.* October 5. https://www.cdc.gov/cpr/readiness/mcm.html.

CDC. 2020b. *Plague: Resources for Clinicians.* February 10. https://www.cdc.gov/plague/healthcare/clinicians.html.

CDC. 2020c. *COVID-19 Vaccination Program Interim Operational Guidance Jurisdiction Operations.* Washington, DC: Department of HHS, October. CDC, Feb. 2020. https://www.cdc.gov/plague/healthcare/clinicians.html.

CDC. 2021a. *Estimated Disease Burden of COVID-19.* January 19. https://www.cdc.gov/coronavirus/2019-ncov/cases-updates/burden.html.

CDC. 2021b. *United States COVID-19 Cases and Deaths by State.* April 2. https://covid.cdc.gov/covid-data-tracker/#cases_casesper100klast7days.

CDC. 2021c. *COVID-19 Vaccine Emergency Use Authorization (EUA) Fact Sheets for Recipients and Caregivers.* January 21. https://www.cdc.gov/vaccines/covid-19/eua/index.html.

CDC. 2021d. *Access and Functional Needs Toolkit: Integrating a Community Partner Network to Inform Risk Communication Strategies.* July 23. https://www.cdc.gov/cpr/readiness/resources/publications/afntoolkit.htm.

CDC. 2021e. *CDC's COVID-19 Vaccine Rollout Recommendations.* March 25. https://www.cdc.gov/coronavirus/2019-ncov/vaccines/recommendations.html.

CDC. 2021f. *How COVID-19 Vaccines Get to You.* March 17. https://www.cdc.gov/coronavirus/2019-ncov/vaccines/distributing.html.

CDC & USDA. 2021. *Federal Select Agent Program.* January 8. https://www.selectagents.gov/.

CDC & USDA. 2022. *Select Agents and Toxins List.* August 30. https://www.selectagents.gov/sat/list.htm.

Coronaviridae Study Group of the International Committee on Taxonomy of Viruses. 2020. "The Species Severe Acute Respiratory Syndrome-related Coronavirus: Classifying 2019-nCoV and Naming it SARS-CoV-2." *Nat Microbiol* 5 (4): 536–544.

DARPA. 2020. *DARPA's Early Investment in COVID-19 Antibody Identification Producing Timely Results.* November 10. https://www.darpa.mil/news-events/2020-11-10.

DARPA. 2023a. *About DARPA.* January. https://www.darpa.mil/about-us/about-darpa.

DARPA. 2023b. *Pandemic Prevention Platform (P3).* January. https://www.darpa.mil/program/pandemic-prevention-platform.

Department of HHS. 2017. *Generating Antibiotic Incentives Now.* Washington, D.C.: Department of HHS. https://www.fda.gov/files/about%20fda/published/Report-to-Congress-on-Generating-Antibiotic-Incentives-Now-%28GAIN%29.pdf.

Department of HHS. 2019. *Public Health Emergency Medical Countermeasures Enterprise Multiyear Budget Fiscal Years 2018-2022.* December. https://aspr.hhs.gov/PHEMCE/Documents/phemce-myb-30Jan2020-508.pdf.

Department of HHS. 2020a. "Declaration Under the Public Readiness and Emergency Preparedness Act for Medical Countermeasures Against COVID-19." *Federal Register* 85: 15198–15203.

Department of HHS. 2020b. *From the Factory to the Frontlines.* Washington, DC: Department of HHS. https://www.hhs.gov/sites/default/files/strategy-for-distributing-covid-19-vaccine.pdf.

Department of HHS. 2020c. *Trump Administration Producing Supply Kits to Safely Administer COVID-19 Vaccines to Americans.* October 30. https://www.hhs.gov/about/news/2020/10/30/trump-administration-producing-supply-kits-safely-administer-covid-19-vaccines-americans.html.

Department of HHS. 2020d. "Determination of Public Health Emergency." *Federal Register* 7316–7317,Vol. 85.

Department of HHS. 2021a. *DRIVe.* https://drive.hhs.gov/.

Department of HHS. 2021b. *COVID-19 Medical Countermeasure Portfolio.* https://www.medicalcountermeasures.gov/app/barda/coronavirus/COVID19.aspx.

Department of HHS. 2021c. *Strategic National Stockpile (SNS).* March 10. https://www.remm.nlm.gov/sns.htm.

Department of HHS. 2021d. *Topic Collection: Mass Distribution and Dispensing of Medical Countermeasures.* April 6. https://asprtracie.hhs.gov/technical-resources/67/mass-distribution-and-dispensing-of-medical-countermeasures.

Department of HHS. 2022. *PHEMCE Partners - Centers for Disease Control and Prevention.* January. https://medicalcountermeasures.gov/phemce/.

Department of HHS. 2023a. *PHEMCE PARTNERS - Administration for Strategic Preparedness and Response (ASPR).* https://medicalcountermeasures.gov/phemce/.

Department of HHS. 2023b. *PHEMCE PARTNERS.* https://medicalcountermeasures.gov/phemce/.

Department of HHS. 2023c. *PHEMCE PARTNERS - The National Institutes of Health (NIH).* January. https://medicalcountermeasures.gov/phemce/.

Department of HHS. 2023d. *PHEMCE PARTNERS - Department of Defense (DOD).* January. https://medicalcountermeasures.gov/phemce/.

Department of HSS. 2023e. *PHEMCE Partners - Food and Drug Administration (FDA).* January. https://medicalcountermeasures.gov/phemce/.

DTRA. 2021. *About DTRA.* https://www.dtra.mil/WhoWeAre/.

DTRA. 2023. *About DTRA.* January. https://www.dtra.mil/WhoWeAre/.

Elusys Therapeutics Inc. 2019. "Prescribing Information for ANTHIM® (obiltoxaximab)." November. https://www.accessdata.fda.gov/drugsatfda_docs/label/2016/125509lbl.pdf.

Emergent BioSolutions Canada Inc. 2020. "Prescribing Information for BAT® [Botulism Antitoxin Heptavalent]." October. https://www.fda.gov/media/85514/download.

Emergent Manufacturing Operations Baltimore LLC. 2018. "Prescribing Information for ANTHRASIL [Anthrax Immune Globulin Intravenous (Human)]." May. https://www.fda.gov/media/91577/download.

Emergent Manufacturing Operations Baltimore LLC. 2021. "Prescribing Information for RAXIBACUMAB." May. https://www.accessdata.fda.gov/drugsatfda_docs/label/2012/125349s000lbl.pdf.

EOP. 2009. "Proclamation 8443 of October 23, 2009; Declaration of a National Emergency with Respect to the 2009 H1N1 Influenza Pandemic." *Federal Register* 74: 55439–55440. https://www.federalregister.gov/documents/2009/10/28/E9-26130/declaration-of-a-national-emergency-with-respect-to-the-2009-h1n1-influenza-pandemic.

EOP. 2020. "Continuation of the National Emergency Concerning the Coronavirus Disease 2019 (COVID-19) Pandemic." *Federal Register* 85: 15337–15338.

FDA. 1989. *Botulism Immune Globulin.* https://www.accessdata.fda.gov/scripts/opdlisting/oopd/detailedIndex.cfm?cfgridkey=34488.

FDA. 2005a. "Authorization of Emergency Use of Anthrax Vaccine Adsorbed for Prevention of Inhalation Anthrax by Individuals at Heightened Risk of Exposure Due to Attack With Anthrax; Availability." *Federal Register*: Volume 70, 5452–5456.

FDA. 2005b. *How to Comply with the Pediatric Research Equity Act.* September. https://www.fda.gov/media/72274/download.

FDA. 2011. *Botulism Antitoxin Heptavalent (A, B, C, D, E, F, G) (Equine).* https://www.accessdata.fda.gov/scripts/opdlisting/oopd/detailedIndex.cfm?cfgridkey=345011.

FDA. 2014. *Expedited Programs for Serious Conditions––Drugs and Biologics Guidance for Industry.* May. https://www.fda.gov/media/86377/download.

FDA. 2015a. *Product Development Under the Animal Rule, Guidance for Industry.* Washington, DC: U.S. Department of HHS, November. https://www.fda.gov/media/88625/download.

FDA. 2015b. *Summary Basis for Regulatory Action - BioThrax.* Washington, DC: Department of HHS. https://wayback.archive-it.org/7993/20170722071225/https://www.fda.gov/BiologicsBloodVaccines/Vaccines/ApprovedProducts/ucm133822.htm#:~:text=Summary%20Basis%20for%20Regulatory%20Action%20-%20BioThrax%C2%A0(PDF%20-%20159KB).

FDA. 2016. *Expanded Access to Investigational Drugs for Treatment Use - Questions and Answers: Guidance for Industry.* June. FDA, Silver Spring.

FDA. 2017. *Emergency Use Authorization of Medical Products and Related Authorities.* Guidance for Industry and Other Stakeholders. January, Silver Spring.

FDA. 2017a. *Emergency Use Authorization of Medical Products and Related Authorities Guidance for Industry and Other Stakeholders.* Washington, DC: Department of HHS, January. https://www.fda.gov/media/97321/download.

FDA. 2017b. *EUA Authorization of Rafa Laboratories Ltd. Atropine Auto-Injector.* April 11. https://www.fda.gov/media/104550/download.

FDA. 2018a. *Pandemic and All-Hazards Preparedness Reauthorization Act of 2013.* November 8. https://www.fda.gov/emergency-preparedness-and-response/mcm-legal-regulatory-and-policy-framework/pandemic-and-all-hazards-preparedness-reauthorization-act-2013-pahpra.

FDA. 2018b. *Other Laws Affecting FDA.* March 29. https://www.fda.gov/regulatory-information/laws-enforced-fda/other-laws-affecting-fda.

FDA. 2018c. *Regulatory Procedures Manual Chapter 2: FDA Authority.* Washington, DC: Department of HHS. https://www.fda.gov/media/77516/download.

FDA. 2018d. *Code of Federal Regulations - Title 21 - Food and Drugs.* March 22. https://www.fda.gov/medical-devices/medical-device-databases/code-federal-regulations-title-21-food-and-drugs.

FDA. 2018e. *Emergency Use Authorization of Medical Products and Related Authorities.* October 17. https://www.fda.gov/regulatory-information/search-fda-guidance-documents/emergency-use-authorization-medical-products-and-related-authorities.

FDA. 2018f. *Expanded Access: Criteria for an Investigational Product to be Made Available for Widespread Use Under a Treatment Protocol or Treatment IND/ Treatment IDE.* January 4. https://www.fda.gov/news-events/expanded-access/expanded-access-criteria-investigational-product-be-made-available-widespread-use-under-treatment.

FDA. 2018g. *Expanded Access Categories for Drugs (Including Biologics).* January 4. https://www.fda.gov/news-events/expanded-access/expanded-access-categories-drugs-including-biologics.

FDA. 2018h. *What We Do.* March 28. https://www.fda.gov/about-fda/what-we-do#:~:text=Freedom%20of%20Information-,FDA%20Mission,and%20products%20that%20emit%20radiation.

FDA. 2018i. *Transfer of Therapeutic Biological Products to the Center for Drug Evaluation and Research.* February 16. https://www.fda.gov/combination-products/jurisdictional-information/transfer-therapeutic-biological-products-center-drug-evaluation-and-research#:~:text=When%20a%20product's%20primary%20mode,will%20be%20assigned%20to%20CBER.

FDA. 2018j. *Developing Products for Rare Diseases & Conditions.* December 20. https://www.fda.gov/industry/developing-products-rare-diseases-conditions.

FDA. 2018k. *Orphan Drug Act - Relevant Excerpts.* March 9. https://www.fda.gov/industry/designating-orphan-product-drugs-and-biological-products/orphan-drug-act-relevant-excerpts.

FDA. 2018l. *Material Threat Medical Countermeasure Priority Review Vouchers - Draft Guidance for Industry.* Washington, DC: Department of HHS. https://www.fda.gov/regulatory-information/search-fda-guidance-documents/material-threat-medical-countermeasure-priority-review-vouchers-draft-guidance-industry.

FDA. 2018m. *Breakthrough Devices Program -- Guidance for Industry and Food and Drug Administration Staff.* Washington, DC: Department of HHS. https://www.fda.gov/media/108135/download.

FDA. 2018n. *Priority Review.* January 4. https://www.fda.gov/patients/fast-track-breakthrough-therapy-accelerated-approval-priority-review/priority-review.

FDA. 2018o. *FDA Approves The First Drug with an Indication for Treatment of Smallpox.* July 13. https://www.fda.gov/news-events/press-announcements/fda-approves-first-drug-indication-treatment-smallpox.

FDA. 2019a. *Expanded Access for Medical Devices.* June 21. https://www.fda.gov/medical-devices/investigational-device-exemption-ide/expanded-access-medical-devices.

FDA. 2019b. *Products Approved for Other Bioterrorism Emergencies.* November 19. https://www.fda.gov/drugs/bioterrorism-and-drug-preparedness/products-approved-other-bioterrorism-emergencies.

FDA. 2019c. August 14. https://www.fda.gov/drugs/bioterrorism-and-drug-preparedness/products-approved-anthrax.

FDA. 2019d. *Products Approved for Anthrax.* August 14. https://www.fda.gov/drugs/bioterrorism-and-drug-preparedness/products-approved-anthrax.

FDA. 2019e. *User Fee Billable Biologic Products and Potencies Approved Under Section 351 of the PHS Act.* March. https://www.fda.gov/media/124227/download.

FDA. 2019f. *MCMi Program Update FY 2019.* Washington, DC: Department of HHS. https://www.fda.gov/media/136121/download.

FDA. 2019g. *CBER Advanced Technologies Team (CATT).* June 27. https://www.fda.gov/vaccines-blood-biologics/industry-biologics/cber-advanced-technologies-team-catt.

FDA. 2019h. *Formal Meetings Between the FDA and Sponsors or Applicants of PDUFA Products Guidance for Industry.* October 18. https://www.fda.gov/regulatory-information/search-fda-guidance-documents/formal-meetings-between-fda-and-sponsors-or-applicants-pdufa-products-guidance-industry.

FDA. 2019i. “Expedited Programs for Regenerative Medicine Therapies for Serious Conditions, Guidance for Industry.” (Department of HHS). https://www.fda.gov/media/120267/download.

FDA. 2020a. *21st Century Cures Act: MCM-Related Cures Provisions.* October 22. https://www.fda.gov/emergency-preparedness-and-response/mcm-legal-regulatory-and-policy-framework/21st-century-cures-act-mcm-related-cures-provisions.

FDA. 2020b. *Regulated Products.* December 15. https://www.fda.gov/industry/import-basics/regulated-products#:~:text=The%20FDA%20regulates%20a%20wide,human%20use%3B%20radiation%2Demitting%20electronic.

FDA. 2020c. *Letter Revoking EUA for Chloroquine Phosphate and Hydroxychloroquine Sulfate; June 15, 2020.* Washington, DC: Deaprtment of HHS. https://www.fda.gov/media/138945/download.

FDA. 2020d. *214787Orig1s000 Summary Review.* Washington, DC: Department of HHS.

FDA. 2020e. *Veklury (remdesivir) EUA Letter of Authorization; Archived October 22, 2020.* Washington, DC: Department of HHS. https://www.fda.gov/media/143188/download.

FDA. 2020f. *Expiration Dating Extension.* August 11. https://www.fda.gov/emergency-preparedness-and-response/mcm-legal-regulatory-and-policy-framework/expiration-dating-extension#expiration.

FDA. 2020g. *What are Medical Countermeasures?* June 1. https://www.fda.gov/emergency-preparedness-and-response/about-mcmi/what-are-medical-countermeasures.

FDA. 2020h. *Guidance and Other Information of Special Interest to MCM Stakeholders.* November 10. https://www.fda.gov/emergency-preparedness-and-response/mcm-legal-regulatory-and-policy-framework/guidance-and-other-information-special-interest-mcm-stakeholders.

FDA. 2020i. *21st Century Cures Act.* January 31. https://www.fda.gov/regulatory-information/selected-amendments-fdc-act/21st-century-cures-act.

FDA. 2020j. *FDA Approves First Treatment for Ebola Virus.* October 15. https://www.fda.gov/news-events/press-announcements/fda-approves-first-treatment-ebola-virus.

FDA. 2020k. *Letter of Authorization - Chloroquine Phosphate and Hydroxychloroquine Sulfate.* March 28. https://www.fda.gov/media/136534/download.

FDA. 2020l. “Enforcement Policy for Ventilators and Accessories and Other Respiratory Devices During the Coronavirus Disease 2019 (COVID-19) Public Health Emergency.” (Department of HHS). https://www.fda.gov/regulatory-information/search-fda-guidance-documents/enforcement-policy-ventilators-and-accessories-and-other-respiratory-devices-during-coronavirus.

FDA. 2021a. *Emergency Use Authorization.* April 1. https://www.fda.gov/emergency-preparedness-and-response/mcm-legal-regulatory-and-policy-framework/emergency-use-authorization#nerveagents.

FDA. 2021b. *CDER's Coronavirus Aid, Relief, and Economic Security Act (CARES Act) Drug Shortage Mitigation Efforts.* February 4. https://www.fda.gov/drugs/drug-shortages/cders-coronavirus-aid-relief-and-economic-security-act-cares-act-drug-shortage-mitigation-efforts#:~:text=In%20addition%20to%20the%20COVID,visibility%20into%20drug%20supply%20chains.

FDA. 2021c. *Guidance & Regulation (Food and Dietary Supplements).* February 3. https://www.fda.gov/food/guidance-regulation-food-and-dietary-supplements.

FDA. 2021d. *Recommendations for Investigational COVID-19 Convalescent Plasma.* February 12. https://www.fda.gov/vaccines-blood-biologics/investigational-new-drug-ind-or-device-exemption-ide-process-cber/recommendations-investigational-covid-19-convalescent-plasma.

FDA. 2021e. *Ebola Preparedness and Response Updates from FDA*. January 13. https://www.fda.gov/emergency-preparedness-and-response/mcm-issues/ebola-preparedness-and-response-updates-fda.

FDA. 2021f. *Justification of Estimates for Appropriation Committees*. https://www.fda.gov/media/135078/download.

FDA. 2021g. *FDA Updates on Hand Sanitizers Consumers Should not Use*. July 29. https://www.fda.gov/drugs/drug-safety-and-availability/fda-updates-hand-sanitizers-consumers-should-not-use#products.

FDA. 2021h. *MCMi Program Update FY 2021*. Washington, DC: Department of HHS.

FDA. 2021i. *About CDER Small Business and Industry Assistance (SBIA)*. March 26. https://www.fda.gov/drugs/cder-small-business-industry-assistance-sbia/about-cder-small-business-and-industry-assistance-sbia.

FDA. 2021j. *COVID-19-Related Guidance Documents for Industry, FDA Staff, and Other Stakeholders*. April 5. https://www.fda.gov/emergency-preparedness-and-response/coronavirus-disease-2019-covid-19/covid-19-related-guidance-documents-industry-fda-staff-and-other-stakeholders.

FDA. 2021k. *CDER Drug and Biologic Accelerated Approvals Based on a Surrogate Endpoint*. Washington, DC: Department of HHS. https://www.fda.gov/drugs/nda-and-bla-approvals/accelerated-approvals#:~:text=CDER%20Drug%20and%20Biologic%20Accelerated%20Approvals%20Based%20on%20a%20Surrogate%20Endpoint.

FDA. 2021l. *Breakthrough Devices Program*. January 5. https://www.fda.gov/medical-devices/how-study-and-market-your-device/breakthrough-devices-program.

FDA. 2021m. *CBER Regenerative Medicine Advanced Therapy (RMAT) Approvals*. April 5. https://www.fda.gov/vaccines-blood-biologics/cellular-gene-therapy-products/cber-regenerative-medicine-advanced-therapy-rmat-approvals.

FDA. 2021n. *CDER Drug and Biologic Animal Rule Approvals*. Washington, DC: Department of HHS.

FDA. 2021o. *Bamlanivimab EUA Letter of Authorization*. Washington, DC: Department of HHS. https://www.fda.gov/media/143602/download.

FDA. 2021p. *Emergency Use Authorization--Archived Information*. March 31. https://www.fda.gov/emergency-preparedness-and-response/mcm-legal-regulatory-and-policy-framework/emergency-use-authorization-archived-information.

FDA. 2022a. *Emergency Use Authorization--Archived Information*. August 10. https://www.fda.gov/emergency-preparedness-and-response/mcm-legal-regulatory-and-policy-framework/emergency-use-authorization-archived-information.

FDA. 2022b. *Pfizer-BioNTech COVID-19 Vaccines*. December 22. https://www.fda.gov/emergency-preparedness-and-response/coronavirus-disease-2019-covid-19/pfizer-biontech-covid-19-vaccines.

FDA. 2022c. *Moderna COVID-19 Vaccines*. January 15. https://www.fda.gov/emergency-preparedness-and-response/coronavirus-disease-2019-covid-19/moderna-covid-19-vaccines.

FDA. 2022d. *21st Century Cures Act: MCM-Related Cures Provisions*. October 28. https://www.fda.gov/emergency-preparedness-and-response/mcm-legal-regulatory-and-policy-framework/21st-century-cures-act-mcm-related-cures-provisions.

FDA. 2022e. *Emergency Use Authorization--Archived Information*. August 10. https://www.fda.gov/emergency-preparedness-and-response/mcm-legal-regulatory-and-policy-framework/emergency-use-authorization-archived-information.

FDA. 2022f. *Expanded Access | Information for Physicians*. November 7. https://www.fda.gov/news-events/expanded-access/expanded-access-information-physicians.

FDA. 2022g. *About CDER Small Business and Industry Assistance (SBIA).* February 8. https://www.fda.gov/drugs/cder-small-business-industry-assistance-sbia/about-cder-small-business-and-industry-assistance-sbia#:~:text=CDER%20SBIA%20is%20often%20the,understanding%20human%20drug%20product%20regulation.

FDA. 2022h. *Transfer of Therapeutic Biological Products to the Center for Drug Evaluation and Research.* March 7. https://www.fda.gov/combination-products/jurisdictional-information/transfer-therapeutic-biological-products-center-drug-evaluation-and-research#1.

FDA. 2022i. *Guidances.* January 24. https://www.fda.gov/industry/fda-basics-industry/guidances.

FDA. 2022j. *Prescription Drug User Fee Amendments.* October 27. https://www.fda.gov/industry/fda-user-fee-programs/prescription-drug-user-fee-amendments.

FDA. 2022k. *Humanitarian Device Exemption.* October 3. https://www.fda.gov/medical-devices/premarket-submissions-selecting-and-preparing-correct-submission/humanitarian-device-exemption.

FDA. 2022l. *Getting a Humanitarian Use Device to Market.* October 3. https://www.fda.gov/medical-devices/humanitarian-device-exemption/getting-humanitarian-use-device-market.

FDA. 2022m. *Breakthrough Devices Program.* October 28. https://www.fda.gov/medical-devices/how-study-and-market-your-device/breakthrough-devices-program.

FDA. 2022n. *Drug and Biologic Accelerated Approvals Based on a Surrogate Endpoint.* September 30. https://www.fda.gov/media/151146/download.

FDA. 2022o. *What are Medical Countermeasures?* December 16. https://www.fda.gov/emergency-preparedness-and-response/about-mcmi/what-are-medical-countermeasures.

FDA. 2022p. *Medical Countermeasure Monitoring and Assessment.* December 16. https://www.fda.gov/emergency-preparedness-and-response/mcm-issues/medical-countermeasure-monitoring-and-assessment.

FDA. 2023a. *In Vitro Diagnostics EUAs.* January 12. https://www.fda.gov/medical-devices/coronavirus-disease-2019-covid-19-emergency-use-authorizations-medical-devices/in-vitro-diagnostics-euas.

FDA. 2023b. *Emergency Use Authorization.* January 13. https://www.fda.gov/emergency-preparedness-and-response/mcm-legal-regulatory-and-policy-framework/emergency-use-authorization.

FDA. 2023c. *Regenerative Medicine Advanced Therapy Designation.* January 10. https://www.fda.gov/vaccines-blood-biologics/cellular-gene-therapy-products/regenerative-medicine-advanced-therapy-designation.

FDA. 2023d. *Cumulative CBER Regenerative Medicine Advanced Therapy (RMAT) Designation Requests Received by Fiscal Year.* January 10. https://www.fda.gov/vaccines-blood-biologics/cellular-gene-therapy-products/cumulative-cber-regenerative-medicine-advanced-therapy-rmat-designation-requests-received-fiscal.

FDA. 2023e. *PDUFA VII: Fiscal Years 2023 – 2027.* January 20. https://www.fda.gov/industry/prescription-drug-user-fee-amendments/pdufa-vii-fiscal-years-2023-2027.

Forbes. 2020. *FDA 'Fast Tracks' First Coronavirus Vaccine From Moderna.* May 12. https://www.forbes.com/sites/alexandrasternlicht/2020/05/12/fda-fast-tracks-first-coronavirus-vaccine-from-moderna/?sh=3dc20d6542aa.

Gottron, F. 2009. *Project BioShield: Purposes and Authorities.* Washington, DC: Congressional Research Service.

ID Biomedical Corporation of Quebec. 2013. "Prescribing Information for Influenza A (H5N1) Virus Monovalent Vaccine, Adjuvanted." November. https://www.fda.gov/media/87479/download.

Innovation Pharmaceuticals, Inc. 2021. *BRILACIDIN.* http://www.ipharminc.com/brilacidin-1.

Janssen Pharmaceuticals, Inc. 2018. "Prescribing Information for LEVAQUIN® (levofloxacin)." July. https://www.accessdata.fda.gov/drugsatfda_docs/label/2018/020634s069lbl.pdf.

JPEO-CBRND. 2022. *Command Brief.* November 21. https://www.jpeocbrnd.osd.mil/Portals/90/2022_JPEOCBRND_Command%20Brief%2016x9_1.pdf.

JPEO-CBRND. 2023. *About.* January. https://www.jpeocbrnd.osd.mil/Our-Story/About/.

Laney, J., and M. Clark. 2018. *FDA Approval of Anti-Seizure Drug Provides a New Tool for Protecting Americans During a Chemical Attack.* October 24. https://www.jpeocbrnd.osd.mil/news/news-story/2018/10/24/fda-approval-of-anti-seizure-drug-provides-a-new-tool-for-protecting-americans-during-a-chemical-attack.

Liu, J., et al. 2020. "Hydroxychloroquine, a Less Toxic Derivative of Chloroquine, Is Effective in Inhibiting SARS-CoV-2 Infection in Vitro." *Cell Discovery,*6-16. doi: 10.1038/s41421-020-0156-0. PMID: 32194981; PMCID: PMC7078228.

Merck Sharp & Dohme LLC. 2022. "Prescribing Information for ERVEBO®- zaire Ebolavirus (strain kikwit-95) Envelope Glycoprotein Injection, Solution." September. https://www.fda.gov/media/133748/download.

Mulangu, S., et al. 2019. "Controlled Trial of Ebola Virus Disease Therapeutics." *New England Journal of Medicine* 381 (24), December 12: 2293–2303. doi:10.1056/NEJMoa1910993. Epub 2019 Nov 27. PMID: 31774950. *New England Journal of Medicine* 381 (24): 2293–2303.

National Library of Medicine. 2020. *Section 1: Introduction to CBRNE Concepts.* May 4. https://www.nlm.nih.gov/dis_courses/cbrne/01-000.html.

National Research Council. 2011. *Protecting the Frontline in Biodefense Research: The Special Immunizations Program.* Washington, DC: The National Academies Press. doi:doi.org/10.17226/13112.

NIAID. 2020. *COVID-19 Clinical Research.* November 2. https://www.niaid.nih.gov/diseases-conditions/covid-19-clinical-research.

NIH. 2021. *Accelerating Covid-19 Therapeutic Interventions ānd Vaccines (ACTIV).* https://www.nih.gov/research-training/medical-research-initiatives/activ.

Novavax, Inc. 2020. *Novavax COVID-19 Vaccine Granted Fast Track Designation by U.S.* FDA, November 9. https://ir.novavax.com/news-releases/news-release-details/novavax-covid-19-vaccine-granted-fast-track-designation-us-fda.

NRx Pharmaceuticals, Inc. 2023. *ZYESAMI®.* January. https://www.nrxpharma.com/zyesami/.

Office of Public Health Emergency Preparedness. 2006. "Draft HHS Public Health Emergency Medical Countermeasures Enterprise (PHEMCE) Strategy for Chemical, Biological, Radiological and Nuclear (CBRN) Threats 1." *Federal Register* 71: 53097–53102.

Office of the ASPR. 2014. *Requirements Branch.* September 24. https://www.phe.gov/about/OPP/mcsr/Pages/requirements.aspx.

Office of the ASPR. 2015. *PHEMCE Mission Components.* February 27. https://www.phe.gov/Preparedness/mcm/phemce/Pages/mission.aspx.

Office of the ASPR. 2017. *PHEMCE Governance.* February 3. https://www.phe.gov/Preparedness/mcm/phemce/Pages/governance.aspx.

Office of the ASPR. 2019a. *Pandemic and All Hazards Preparedness Act.* September 17. https://www.phe.gov/Preparedness/legal/pahpa/Pages/default.aspx.

Office of the ASPR. 2019b. *PHEMCE Multiyear Budget, Fiscal Years 2018-2022.* Washington, DC: Department of HHS.

Office of the ASPR. 2019c. *Pandemic and All-Hazards Preparedness Reauthorization Act.* September 17. https://www.phe.gov/Preparedness/legal/pahpa/Pages/pahpra.aspx.

Office of the ASPR. 2019d. *Pandemic and All-Hazards Preparedness and Advancing Innovation Act.* September 27. https://www.phe.gov/Preparedness/legal/pahpa/Pages/pahpaia.aspx.

Office of the ASPR. 2020. *Background.* January 29. https://www.phe.gov/Preparedness/mcm/phemce/phemce-myb/FY2018-2022/Pages/background.aspx.

Office of the ASPR. 2021a. *Public Readiness and Emergency Preparedness Act.* March 16. https://www.phe.gov/Preparedness/legal/prepact/Pages/default.aspx.

Office of the ASPR. 2021b. *Sustaining the Stockpile.* February 26. https://www.phe.gov/about/sns/Pages/sustaining.aspx.

Office of the ASPR. 2021c. *Strategic National Stockpile.* February 26. https://www.phe.gov/about/sns/Pages/default.aspx.

Office of the ASPR. 2021d. *Response to the COVID-19 Pandemic.* April 2. https://www.phe.gov/about/sns/COVID/Pages/default.aspx.

Office of the ASPR 2021e. https://aspr.hhs.gov/SNS/Pages/SNS-Role-Federal-Vaccination-Campaign.aspx

Office of the ASPR. 2023a. *Public Health Emergency.* Office of the ASPR Website. January 11. https://aspr.hhs.gov/legal/PHE/Pages/covid19-11Jan23.aspx.

Office of the Assistant Secretary for Preparedness and Response [ASPR]. 2021. *Public Health Emergency.* February 18. https://www.phe.gov/emergency/news/healthactions/phe/Pages/default.aspx.

Patel, K. B., et al. 2018. "Snow DM, Cobb RR, Martinez J, Finger-Baker I, Collins L, Terpening S, Syar ES, Niemuth N, Kobs D, Barnewall R, Farr-Jones S, Marks JD, Tomic MT. A Monoclonal Antibody Combination against both Serotypes A and B Botulinum Toxin Prevents Inhalational Botulis." *Applied Environmental Microbiology* 84 (24): e01280–18. doi:10.1128/AEM.01280-18.

Pfizer Inc. 2020. *Pfizer and BioNTech Granted FDA Fast Track Designation for Two Investigational mRNA-based Vaccine Candidates Against SARS-CoV-2.* July 13. https://www.pfizer.com/news/press-release/press-release-detail/pfizer-and-biontech-granted-fda-fast-track-designation-two.

PHEMCE. 2017. "2017-2018 Public Health Emergency Medical Countermeasures (PHEMCE) Strategy and Implementation Plan." *PHEMCE Website.* December 18. https://www.phe.gov/Preparedness/mcm/phemce/Documents/2017-phemce-sip.pdf.

Philips. 2020. *The U.S. Department of Defense, Philips Expand Research on Groundbreaking AI Technology, Opening New Front in Battle to Contain COVID-19.* September 22. https://www.usa.philips.com/a-w/about/news/archive/standard/news/press/2020/20200922-the-us-department-of-defense-philips-expand-research-on-groundbreaking-ai-technology-opening-new-front-in-battle-to-contain-covid-19.html.

Rafa Laboratories, Ltd. 2019. *Prescribing Information for Atropine Injection.* July. https://www.accessdata.fda.gov/drugsatfda_docs/label/2019/212319s002lbl.pdf.

Rasetti-Escargueil, C., and M. R. Popoff. 2019. "Antibodies and Vaccines against Botulinum Toxins: Available Measures and Novel Approaches." *Toxins (Basel)* 11 (9): 528.

Ridley, D. B., and S. A. Régnier. 2016. "The Commercial Market For Priority Review Vouchers." *Health Aff (Millwood)* 35 (5): 776–783. https://pubmed.ncbi.nlm.nih.gov/27140982/.

Rodríguez-Díaz, C. E., et al. 2018. "Zika Virus Epidemic in Puerto Rico: Health Justice Too Long Delayed." *International Journal of Infectious Diseases* 65: 144–147. doi:10.1016/j.ijid.2017.07.017.

Sanofi Pasteur. 2007. "Prescribing Information for Influenza Virus Vaccine, H5N1." April. https://www.fda.gov/media/74534/download.

Seqirus Inc. 2020. "Prescribing Information for AUDENZ (Influenza A (H5N1) Monovalent Vaccine, Adjuvanted)." February. https://www.fda.gov/media/135020/download.

Sheahan, T. P., et al. 2017. "Broad-spectrum Antiviral GS-5734 Inhibits Both Epidemic and Zoonotic Coronaviruses." *Science Translational Medicine* 9 (396): eaal3653.

Shrestha, S., et al. 2011. "Estimating the Burden of 2009 Pandemic Influenza A (H1N1) in the United States (April 2009-April 2010)." *Clinical Infectious Diseases* (52 Suppl 1) S75–82. doi:10.1093/cid/ciq012.

SIGA Technologies, Inc. 2022. "Prescribing Information for TPOXX® (tecovirimat)." May. https://www.accessdata.fda.gov/drugsatfda_docs/label/2018/208627s000lbl.pdf.

Snow, D. M., et al. 2021. "A Monoclonal Antibody Combination against both Serotypes A and B Botulinum Toxin Prevents Inhalational Botulism in a Guinea Pig Model." *Toxins* 13: 31. doi:10.3390/toxins13010031.

U.S. House of Representatives 2003. https://www.congress.gov/event/108th-congress/house-event/LC14476/text.

U.S. National Library of Medicine. 2021. *Search Results.* April 2. https://www.clinicaltrials.gov/ct2/results?term=covid-19&age_v=&gndr=&type=Expn&rslt=&Search=Apply.

Vazquez-Cintron, E., et al. 2020. "Symptomatic Treatment of Botulism with a Clinically Approved Small Molecule." *JCI Insight* 5 (2): e132891. doi:10.1172/jci.insight.132891.

Vergun, D. 2020. *AI Aids DOD in Early Detection of COVID-19.* September 22. https://www.defense.gov/Explore/News/Article/Article/2356086/ai-aids-dod-in-early-detection-of-covid-19/.

Wang, M., et al. 2020. "Remdesivir and Chloroquine Effectively Inhibit the Recently Emerged Novel Coronavirus (2019-nCoV) in Vitro." *Cell Research* 30: 269–271.

Index

D

E

F

G

H

Made in the USA
Las Vegas, NV
19 September 2026

c914d726-2290-436d-bf14-76b83f6af173R01